Praise for Buzz Aldrin's *Mission to Mars*

"A masterful array of strategies for exploration by a true
space expert and patriot."
—*Michael Collins, astronaut and command module pilot, Apollo 11*

"*Mission to Mars* is pure Buzz: creative ideas flying off the pages,
a big picture view of how to move forward, and a laser-like focus on
why space exploration is key to humanity's future. Buzz Aldrin has been
making major contributions to the U.S. space program for a half century,
and his new book continues that tradition."
—*John M. Logsdon, founder of the Space Policy Institute and professor emeritus,
The George Washington University*

"I've traveled to the North Pole with Buzz, and if I were to travel to Mars I
can't think of a better person to plan the trip than he. Sign me up!"
—*Norman Augustine, Chairman, Review of U.S. Human Space Flight Plans Committee*

"Buzz Aldrin's *Mission to Mars* presents a bold, inviting plan
to colonize Mars. His call that the 'Earth isn't the only world
for us anymore' is incontrovertible."
—*Roger D. Launius, Senior Curator, Division of Space History,
National Air and Space Museum*

"No one's given more thought to Mars exploration than Buzz Aldrin—
a hero whose legacy as one of the first men on the moon may well be eclipsed
by his contributions to engineering our future in space."
—*Elliot Holokauahi Pulham, CEO, Space Foundation*

"Colonizing space is essential for the long-term survival of the human race,
and this book shows us how."
—*Stephen Hawking*

"Any time an Apollo-era astronaut steps forward with ideas for our future
in space, it's time to stop whatever we're doing and pay attention. Buzz Aldrin,
one of the first moonwalkers, has no shortage of these ideas. And in *Mission
to Mars* he treats us to how, when, and why we should travel there."
—*Neil deGrasse Tyson*

"Buzz is one of the foremost forward thinkers of our time, and this book
will be essential reading for those who care about humanity's future in space."
—*Richard Branson*

MISSION TO
MARS

MY VISION for SPACE EXPLORATION

BUZZ ALDRIN

WITH LEONARD DAVID

FOREWORD BY ANDREW ALDRIN

NATIONAL GEOGRAPHIC

WASHINGTON, D.C.

Published by the National Geographic Society
1145 17th Street N.W., Washington, D.C. 20036

ISBN: 978-1-4262-1017-4 (hardcover)
ISBN: 978-1-4262-1468-4 (paperback)
First paperback printing 2015

Library of Congress Preassigned Control Number: 2012953599

The National Geographic Society is one of the world's largest nonprofit scientific and educational organizations. Founded in 1888 to "increase and diffuse geographic knowledge," the Society's mission is to inspire people to care about the planet. It reaches more than 400 million people worldwide each month through its official journal, *National Geographic,* and other magazines; National Geographic Channel; television documentaries; music; radio; films; books; DVDs; maps; exhibitions; live events; school publishing programs; interactive media; and merchandise. National Geographic has funded more than 10,000 scientific research, conservation and exploration projects and supports an education program promoting geographic literacy.

For more information, visit www.nationalgeographic.com.

National Geographic Society
1145 17th Street N.W.
Washington, D.C. 20036-4688 U.S.A.

For information about special discounts for bulk purchases, please contact National Geographic Books Special Sales: ngspecsales@ngs.org

For rights or permissions inquiries, please contact National Geographic Books Subsidiary Rights: ngbookrights@ngs.org

Interior design: Katie Olsen

Printed in the United States of America

15/QGF-CML/1

CONTENTS

✦ ✦ ✦

Andrew Aldrin and his father, Buzz Aldrin, 2012

FOREWORD

BY ANDREW ALDRIN

This book represents a journey of its own. It began the moment my father set foot back on Earth. Since that time he has been constantly thinking about how people from Earth will inhabit another planet. What you will read is a vision of the extension of humanity to Mars, to be sure. But the strength of my father's vision is the pathway as much as it is the planet. It is about the journey as much as the destination.

I was there for most of it. I remember some of it.

My earliest recollections are probably of me sitting around the kitchen table looking at these fantastic models of space planes, and understanding almost nothing about the space shuttle that was going to be the next generation of spacecraft intended to fly people into orbit. There was an impassioned discussion—well, more of a monologue—regarding early designs for a space shuttle system that had a piloted flyback booster for the first stage and why we needed to separate crew from cargo. That is about all I remember. I was 11 and had nothing to contribute, except for a willing ear.

And that is the way it went for the first 20 years or so of our conversations about space travel. I listened, and I learned, while my dad talked about space.

Some of these conversations involved the great thinkers of our time. When my dad came up with an idea, he would seek out the people who were doing the most creative thinking at the moment. They would usually pick up the phone when he called. At one point he began to look critically at the design of the space station. The structure just didn't seem efficient. I recall that he became enamored with geodesic structures, so he naturally called on Buckminster Fuller. Now that was an amazing set of conversations. At the time it seemed like competing soliloquies. But I began to see many more of Bucky's ideas creep into the design.

And it is not just the well known and famous who captured my father's attention.

If anyone had an idea that fit in my dad's vision of the future, he would go and talk to them. Often he became the tireless advocate for their ideas. He was building an architecture of systems. If someone had an element that fit in his structure, he wanted to use it.

You will see a lot of other people's elements in the pages of this book. But the system, in the broadest sense possible, is truly his own.

All of the pieces fit together. That may be the real virtue of this book. There are hundreds, probably thousands, of people who have talked with my father, read his articles, heard his speeches, and watched him on TV. Each of them contributed a snippet to Dad's overall vision. Sometimes the snippets seemed impossibly disjointed. There was just so much he was trying to articulate, and so little time in one conversation, speech, or interview.

But I recall one conversation with Brent Sherwood, a space architect now at the Jet Propulsion Laboratory, who really does see the big picture. My dad had an incomprehensibly complex diagram of how humans could get from where they were to Mars. Reasonably intelligent people could disagree over many of the points in the diagram. And they did. But afterward Brent told me that, despite his technical differences, "it all hangs together." And he went on to say that Buzz may be the only person who has all of the pieces to the puzzle.

This book is the first time my father has attempted to put the entire puzzle together in one place.

My dad has spent a great deal of time conversing about space with many people. This book is a continuation of that tradition. The basis of these chapters began with the 40 years of publications, interviews, and presentations my dad has made. Many of these have been carefully collected by the people who work closely with my dad. A special mention should be made of the work Christina Korp and Rob Varnas have done in collecting and organizing the considerable library of my father's works.

But the real heart of this book is a series of conversations between my dad, Leonard David, and me over the better part of 2012, all dedicated to the writing of this book. These conversations have been wide ranging, incredibly insightful, entertaining, frustrating, enlightening, and hopelessly disorganized. It is a testament to Leonard's perseverance that he has been able to organize the discussions into an entirely coherent manuscript which, I believe, offers the clearest, most comprehensive account of my father's views on spaceflight.

The book begins with a brief account of my father's flight on Air Force One on the day that U.S. President Barack Obama made his only major speech on space. My dad was more than just a passenger that day. For the past year he had been working tirelessly, refining concepts for space exploration programs, making speeches on space exploration, testifying before space commissions and Congress. My proudest moment was watching his speech at the Smithsonian's National Air and Space Museum on the 40th anniversary of the moon landing. That speech laid out succinctly his vision for the future of human space development and exploration.

This book serves to further that vision.

Space exploration has never been a purely technical enterprise for my father. While the lessons of Apollo are legion, none has made a greater impression on him than the power of the commitment of an American President. John F. Kennedy's speech declaring a clear and compelling goal for space exploration forever changed the course of human history. So, in any conversation with him about the long-term future of human exploration, you will inevitably find yourself discussing space in terms of four-year electoral cycles. Much as you must have a progression of technical development, my father recognizes that there must be a progression of political decisions, supported by sound policy analyses.

In the second chapter, he calls for a recommitment to U.S. space leadership via his Unified Space Vision. To support this, he would like to see the formation of a permanent senior executive, nongovernmental advisory group he calls the United Strategic Space Enterprise.

A long time before it became fashionable, my dad was an unstinting advocate for private space travel. In large part this is because he recognizes the importance of creating large, sustainable markets for launches in order to bring down the costs of a single mission. Also, it is because he believes that participation will be the best way of generating broad public interest in an ambitious space program. In chapter 3 you will read how commercial passenger travel should drive the requirements for reusable Earth-to-orbit transportation systems. Another theme that will come out in this chapter is the importance of utilizing existing systems and infrastructure.

Whether it was taking existing launch vehicles and using them as the basis for reusable flyback boosters, or taking the technology developed by the Soviet Union and United States for smaller, more efficient shuttle-like vehicles, or using existing launch vehicles instead of developing new systems with low flight rates, my dad has always used the technology of today to serve as the basis for revolutionary new systems.

Chapter 4 considers the return to the moon. As strange as it may sound, for the 40 years my dad and I have been talking about space, there have been relatively few times that we have talked about the moon. Been there, done that. The moon seemed a distraction from the real goal—Mars. But over the past several years the moon has come back as a critical element in his thinking. It is the place where the transition from development to exploration takes place. The focus of U.S. policy should be on establishing the transportation, infrastructure, and habitation systems around the moon to enable commercial and international development of the moon. The focus of exploration should be on getting to Mars.

In 2008 my dad began to formulate the idea of using missions to near-Earth asteroids as early precursors to Mars missions. This is the focus of chapter 5. I am sure that he was not the first person to consider these missions, but when he combined asteroid missions with the lunar strategy as noted previously, and Phobos as a waypoint to Mars, he created the foundations of the Flexible Path architecture. This approach was later recommended by the Obama-established Augustine Committee as an affordable, progressive path to human exploration.

A key element to the Flexible Path was the concept of landing—docking, really—at Phobos, one of two Mars moons, before proceeding to putting humans on the red planet. I have to admit the logic of this was not immediately obvious to me. Why do you go 99.999 percent of the distance to Mars without completing the mission? Chapter 6 focuses on how a stop at Phobos creates the basis for a more permanent presence on Mars by allowing astronauts to teleoperate the systems that piece together the infrastructure necessary to sustain human habitation of the planet.

My father has done a great many things in his life. The entire world respects and admires him for participating in the first human spaceflight to the moon. But from what I see, for him, innovative technical concepts matter more: developing key elements of the rendezvous approaches that were critical to the success of early human spaceflight missions; revolutionizing extravehicular training through underwater exercises—these were some of his contributions to human spaceflight that really made a difference. But the idea of cycling spacecraft between planetary objects may be the idea of which my father is the most proud. I think we have been discussing cycling ships for more than 20 years, whether they were between Earth and its moon or the much more technically challenging task of cycling between Earth and Mars.

In chapter 7 my father discusses how we can create a sustainable transportation system to ferry people, supplies, and equipment to Mars by utilizing a large host spacecraft on a trajectory constantly traveling between Earth and Mars at regular intervals. This system of cyclers establishes the foundation for the permanent human inhabitation of Mars.

Over the 40 years of conversations about space, I really can't remember a single time that my dad talked to me about his trip to the moon. Sure, there were brief words here and there, but the conversation was always about the future. He cares about where we are going as a civilization, not where we have been. In chapter 8 he concludes with a discussion about what it will take to get us to Mars. His vision is not just about the technical or programmatic elements, it is about the political and social underpinnings necessary to reinvigorate the nation's commitment to the human exploration and development of space. He looks to 2019, the 50th anniversary of his historic landing on the moon, as the date on which the American President will make a commitment to establish a permanent human presence on Mars.

That is the clarion call of this book.

Buzz Aldrin on the moon, July 1969

CHAPTER ONE

THE VIEW FROM AIR FORCE ONE

It was April 15, 2010, when I stepped off Air Force One, one of several guests invited to travel with U.S. President Barack Obama to the John F. Kennedy Space Center at Merritt Island, Florida. I had some inkling that he was going to push the reset button on the U.S. space program.

As he began a formal address, President Obama was kind enough to recognize me among those gathered. It was certainly ego lifting to hear him say: "Four decades ago, Buzz became a legend. But in the four decades since he's also been one of America's leading visionaries and authorities on human spaceflight." Welcomed words, but I patiently awaited the punch lines of his speech.

"Now, I understand that some believe that we should attempt a return to the surface of the Moon first, as previously planned," the President said. "But I just have to say pretty bluntly here . . . we've been there before. Buzz has been there. There's a lot more of space to explore, and a lot more to learn when we do."

Obama then called for a series of increasingly demanding targets, made feasible by advancing our technological capacity with each progressive step forward.

"Early in the next decade, a set of crewed flights will test and prove the systems required for exploration beyond low Earth orbit. And by 2025, we expect new spacecraft designed for long journeys to allow us to begin the first-ever crewed missions beyond the Moon into deep space. So, we'll start . . . we'll start by sending astronauts to an asteroid for the first time in history. By the mid-2030s, I believe we can send humans to orbit Mars and return them safely to Earth. And a landing on Mars will follow. And I expect to be around to see it," the President explained, greeted by bursts of applause throughout the rollout of his prospective missions.

President Obama's rejection of the previous space plan—the Vision for Space Exploration—announced by President George W. Bush in January 2004, was in motion. That Bush space agenda spurred NASA to orchestrate the Constellation Program, one that involved development of two booster vehicles: Ares I and Ares V. Ares I would boost crews into orbit. The Ares V heavy-lift launcher would hurl other hardware into space. Along with these two boosters, the Constellation Program called for a set of other spacecraft to be built, including the Orion crew capsule, an Earth departure stage, and the Altair lunar lander.

Coupled with the retirement of the space shuttle program in 2010, the Bush proposal would establish an extended human

President Obama and Buzz Aldrin together in Florida for the President's space policy speech, April 2010

presence on the moon and begin a "sustainable course of long-term exploration."

As President Obama took the stage, America's civil space program had been aimed at returning astronauts to the moon by 2020. Also, that Bush plan called for cultivating the technologies that would support human expeditions to Mars, our ultimate destination in space.

Still, I was always worried about a phrase Bush used: *a* mission to Mars. My view: You're not going to sell *a* mission to Mars.

In the long run, two things happened along the way following President Bush's outlining of his Vision for Space Exploration. First, Bush failed to fully fund the program, as he had initially promised. As a result, each year the development of the rockets and spacecraft called for in the plan slipped further and further behind. Second, and most important, NASA virtually eliminated the technology development effort for advanced space systems. Equally as bad, NASA also raided the Earth and space science budgets in the struggle to keep the program then named Project Constellation on track. Even that effort fell short. To keep the focus on the return to the moon, NASA pretty much abandoned all hope of preparing for Mars exploration with humans.

Obama's dismissal of the Bush Vision for Space Exploration was fortified by the earlier findings of a panel of space experts, which concluded that the Constellation Program could not be executed without very substantial increases in funding. That panel was headed up by my friend Norm Augustine, past head of Lockheed Martin and a former government official. Augustine's team had received testimony and presentations from across the space community, including mine in 2009, as to how NASA should shape its future. In the end, the Augustine Committee observed that the path established by Bush was not sustainable, and President Obama agreed. Augustine and I both zeroed in on the truth that the Bush vision was unattainable and the country needed something else.

The option clearly was a flexible path, somehow. At the time I felt that we had an option of extending space shuttle flights, perhaps developing a shuttle-derived capability. But soon it became clear that extending the shuttle program was not an economically viable thing to do, so now we have a gap in America's independent access to space.

I did believe the Bush vision for space was a good, albeit flawed, notion. It moved away from the space shuttle and the International Space Station and back to exploration, *somewhere*—even though back to the moon with government astronauts was not to my liking. I did concur that Constellation required extensive reevaluation. Obama's action to cancel Constellation, however, has morphed into the Space Launch System (little more than the canceled Ares V booster in the Constellation Program) and Orion, which is ill named as a multipurpose crew vehicle. Why? Overall, it is because of short-term, vested interests in political and industrial circles. It's my opinion that some of the large aerospace contractors are far from truthful in working with NASA.

At the moment NASA's own website on Constellation tells the story as an editorial note: "The Constellation program is no longer an active NASA program. The program information on these pages is for historical use only."

So be it for historical artifacts. But first a little history about my own space journey to today.

Beyond the Boundaries

I know firsthand that challenging times often come first before the most rewarding moments. Over the centuries we have seen powerful reminders of those who explored beyond the boundaries of what they knew, from Copernicus and Galileo to Columbus. Jumping to the 20th century, it was on a windswept morning in 1903 at Kitty Hawk that the Wright brothers made the first powered flight. That same year, my mother, Marion Moon, was born.

My father, Edwin Eugene Aldrin, was an engineer and an aviation pioneer—and a friend of Charles Lindbergh and Orville Wright. Taking a job with Standard Oil, my dad flew his own plane coast to coast. He later served in World War II in the Army Air Corps, coming home for visits.

Born in 1930 and raised in Montclair, New Jersey, I finished high school there. Aviation was pretty much in the family. When I was all of two years of age, my dad took me on my first flight, the two of us winging our way from Newark down to Miami to visit relatives. My aunt, in fact, was a stewardess for Eastern Airlines. The Lockheed Vega single-engine plane that I flew in was trimmed in red paint to look like an eagle. How could I have grasped then as a child that decades later I would find myself strapped inside a very different breed of flying machine—Apollo 11's lander, the *Eagle,* en route to the moon's Sea of Tranquillity?

Aldrin family holiday card signed by all, including "Buzzer"

The heritage that led me into aviation and the appreciation for higher education came from my father. Dad had gone to Clark University in Worcester, Massachusetts. His physics professor was Robert Goddard, regarded as the father of liquid-fueled rocketry.

After graduation from high school, I became a cadet at West Point and took to heart its motto, "Duty, Honor, Country." It's a maxim that remains part of me today. Surrounded by the influence of aviation, I entered the U.S. Air Force after graduating from the Military Academy. After fighter pilot training I was stationed in Korea, where I flew 66 combat missions in my F-86 Sabre fighter jet, shooting down two enemy MiG-15 aircraft.

Following the Korean War, I was sent to Germany and was on alert, flying F-100s that carried nuclear weapons. In the late

Buzz climbs into his F-86 Sabre Jet in Korea, circa 1952.

1950s the Cold War was escalating between the then Soviet Union and the United States. To be sure, tensions were high. While posted in Germany, I learned of the Soviets' surprising technological feat—the launch of Earth's first artificial satellite in October 1957, a 184-pound sphere called Sputnik. As the import of Sputnik sank in, against the backdrop of the Cold War, the political and public reaction spurred on the space age. It became the starting gun for the space race, leading to the creation of NASA the following year.

The Soviet Union achieved yet another triumph on April 12, 1961, by sending the first human into Earth orbit, cosmonaut Yuri Gagarin, in his Vostok 1 spacecraft. As a comparative note, a few weeks after Gagarin's mission of 108 minutes duration, NASA flew on May 5 America's first Mercury astronaut, Alan Shepard, on a 15-minute suborbital flight that touched the edge of space.

A mere 20 days after Shepard's mission, President John F. Kennedy boldly challenged America to commit itself to achieving the goal of landing a man on the moon before the end of that decade. Many of those at the helm of a newly formed NASA thought the challenge to be impossible. The know-how just wasn't there. The nation had little more than 15 minutes of spaceflight experience under its belt.

But what America *did* have was a President with vision, determination, and the confidence that such a goal was attainable. By publicly stating our goal and by establishing an explicit time period on a very clear accomplishment, President Kennedy offered no back door. We either had to do it or not make the grade . . . and no one was interested in failing. Even then, failure was not an option.

Kennedy's audacious objective was further reinforced by his speech at Rice University on September 12, 1962. That seminal speech included the famed line: "We choose to go to the moon in this decade and do the other things, not because they are easy, but because they are hard." That presentation, even today, remains riveting.

Kennedy's empowering words from over 50 years ago are worth recalling in terms of the technical challenges we face today. In part, he said,

we shall send to the moon, 240,000 miles away from the control station in Houston, a giant rocket more than 300 feet tall . . . made of new metal alloys, some of which have not yet been invented, capable of standing heat and stresses several times more than have ever been experienced, fitted together with a precision better than the finest watch, carrying all the equipment needed for propulsion, guidance, control, communications, food and survival, on an untried mission, to an unknown celestial body, and then return it safely to earth, re-entering the atmosphere at speeds of over 25,000 miles per hour, causing heat about half that of the temperature of the Sun . . . and do all this, and do it right, and do it first before this decade is out—then we must be bold.

Rendezvous With Destiny

If space was going to be our next new frontier, then I wanted to be part of getting there. After completing my tour of duty

The Huntsville Times

Feature Index

24 PAGES TODAY

VOL. 51, NO. 21   CHICAGO DAILY NEWS SERVICE   HUNTSVILLE, ALABAMA, WEDNESDAY, APR. 12, 1961   ASSOCIATED PRESS — WIREPHOTO   45c PER WEEK

Where Progress...

Covers The Valley!

Man Enters Space

'So Close, Yet So Far,' Sighs Cape

U.S. Had Hoped For Own Launch

CAPE CANAVERAL, Fla. (AP) — The Redstone rocket which the United States had hoped would boost the first man into space stands on a launching pad here. The Soviet Union beat its firing date by at least two weeks.

"So close, yet so far," commented a technician who is helping groom the Redstone to send one of America's astronauts on a short sub-orbital flight, hopefully late this month or early in May.

"If we hadn't had those troubles last fall and on the shroud and Little Joe shots this year, we might have made it," the technician said.

"But you have to give the Russians scientific credit. They've accomplished a remarkable breakthrough."

Dr. Hugh Dryden, deputy director of the National Aeronautics and Space Administration, told Washington Tuesday that the earliest possible date for the manned launching is about April 26.

Project Mercury officials hoped to advance a manned Redstone flight last December or January. A series of launch mishaps necessitated additional launchings to qualify the system.

On Nov. 8, a space capsule failed to separate from a Little Joe booster from Wallops Island, Va., in a test of the escape system.

Two weeks later, a Redstone failed because of a faulty connection which raised the escape tower to fire, leaving the rocket and capsule on the pad. This failed to be repeated before March, the same circumstance was seen on a short trip Jan. 31.

An engine thrust regulator stuck on the chimp shot, creating an excessive thrust which lofted the chimp, Ham, higher and farther than intended. Another Redstone was fired to press out corrections made to the regulator, again delaying the manned trip.

Another setback occurred in March 18 when a repeat of the Jan. 31 shot delayed the test. Friday on three tries found the difficulty. Now of course there is the one.

By "this one," he referred to the Moscow announcement that Maj. Yuri A. Gagarin made a successful flight around the earth in a five to seven-ton space ship and returned safely.

'Worker' Stands By Story

LONDON (AP) — The Daily Worker, Communist party paper in Britain, said today it in fact may lay by its story that the Soviet Union launched a man into space and had high finished.

"Our story came from good sources. All we know is what we published today. Now of course there is this one."

By "this one," he referred to the Moscow announcement that Maj. Yuri A. Gagarin made a successful flight around the earth in a five to seven-ton space ship and returned safely.

This is Russian Maj. Yuri Gagarin, history's first man in space. The Russians today rocketed him around the earth in an orbit taking slightly less than 90 minutes and brought him back safely to a prearranged spot in the Soviet Union. (AP Wirephoto via radio from Moscow)

Praise Is Heaped On Major Gagarin

First Man To Enter Space Is 27, Married, Father Of Two

LONDON (AP) — The Daily Worker, Communist party paper in Britain, said today it in fact may lay by its story that the Soviet Union launched a man into space and had high finished.

"For those who did not see this picture we should like to give a description of this splendid man."

"On the screen appears the image of a man aged about 30 with a kind, Russian face, open and well aged. The body brave and high forehead."

The portrait of Maj. Yuri A. Gagarin was shown and then came this broadcast comment, repeated by Moscow radio.

Soviet Officer Orbits Globe In 5-Ton Ship

Maximum Height Reached Reported As 188 Miles

MOSCOW (AP)—A Soviet astronaut has orbited the globe for more than an hour and returned safely to receive the plaudits of scientists and political leaders alike, Soviet announcement of the feat brought praise from President Kennedy and U.S. space experts left behind in the contest to put the first man into successful space flight.

By the Soviet account, Maj. Yuri Alekseyevich Gagarin, gade a five-ton spaceship once around the earth in an orbit taking an hour and 29 minutes. He was in the air a total of an hour and 48 minutes.

The whole sequence of events and the announcements relating to it raised a number of questions. The Soviet announcement said the flight took place today between 9:07 and 10:55 a.m., but some persons in Moscow's Western colony were skeptical that the feat actually come off today.

Rumors had been circulating several days that the space craft had been poised off. Two days ago, Soviet TV technician raced the Central Telegraph Office with the evident purpose of getting pictures of correspondents in action as they reported such a story. There was various reports, some veritable from official reports, that the flight had been made.

Two Tuesday night the Daily Worker, London Communist newspaper with apparently usual connections in Moscow, reported that the flight took place last Friday in quick snatches, the Daily Worker heralded "the first man in space," saying he had rocketed three orbits before returning to earth resulting from a suffering from "after-effects of the flight."

The full led up to today. About 9:30 a.m., Western correspondents were tipped off to be listening to their radios at 10 a.m. The announcement came at 10 a.m., saying the astronaut and was in fine shape. It was delivered in the radio broadcast messages, reportedly from him over South America and Africa.

Then came the announcement that the spacecraft had been racked back to earth.

VON BRAUN'S REACTION:

To Keep Up, U.S.A. Must Run Like Hell'

Victoria by William McDermott

WERNHER VON BRAUN
Dr Praises A Russian Achievement

By BILL AUSTIN
Of The Times Staff

A disappointed Dr. Wernher von Braun, arriving in Huntsville today, called Russia's space flight a tremendous thing and lauded it the "shot heard around the world."

"I'm disappointed because two again we came in in second place.

"You Braun arrived at the Huntsville airport from Green City, Pa, where he had addressed a military group yesterday.

He said we had hoped all along the United States would be the first to place an astronaut in orbit, but added Russia did an excellent job in history if anything had gone wrong the first American space flight once again the Soviet satellite continued to the first American astronaut.

"We are going to have to run like a race or order, but it is not cut too far behind," he continued.

He also said we can continue whether the Laurens Daily Worker story's first attempt to make a what or not, but the real Russia's first attempt is the "shot heard around the world."

Reds Deny Spacemen Have Died

By THE ASSOCIATED PRESS

Have some Soviet astronauts been killed in space flight experiments before Yuri A. Gagarin's successful trip?

U.S. Soviet officials insist, and some Western sources say they believe no or a low Russians did perish in unannounced attempts. Brig. Gen. Don Flickinger, head of the medical studies of the U.S. Air Force astronaut selection and training program, says he thinks

Turn to Page 2, Column 1

No Astronaut Signal Received At Ft. Monmouth

FORT MONMOUTH, N.J. (AP)— The Army Signal Corps did not receive any radio signals from the Soviet satellite continuing the astronaut it was learned today.

"We are going to have to be the left to catch up," he continued.

He was told we can continue whether the Russia's first attempt is the "shot heard around the world."

The center, operated by the U.S. Army Signal Engineering Laboratories, attempted to monitor radio transmissions from outer space during official would not be commenting on this to the instruments he would be sending any state to use over Fort Monmouth, New Jersey.

When he added message of the nations transmissions to the United States will have addressed a military group in place an astronaut in first, be said Russia has an excellent job in history of anything had gone particularly noteworthy.

"We are going to have to run like a race or order, but it is not cut too far behind," he continued.

Hobbs Admits 1944 Slaying

By DON WARD
Of The Times Staff

Ishare Is Hobbs confessed today to the brutal murder in 1944 of Mrs. Margaret Thomas Florida, Circuit Solicitor Maton L. Weaver said.

Hobbs, now 41 is held by Air Force authorities at Eglin Air Force Base, Fla. He verbal statement there detailing his hatchet-slaying of the prominent 33-year-old widow, Weaver said he learned from air force officials.

The suspect, who has undergone psychiatric treatment by military authorities since Feb. 4, has recounted his memory to Lt. Weaver was told. Psychiatrists and Hobbs' apparent amnesia resulted from "hysteria" rather than from any medical cause.

Hobbs, who attempted suicide last November at Bartow, Fla., and was then exposed as a longtime fugitive, will be returned to Marfield Air Force Base, near Tampa, Fla., from Eglin AFB, Weaver plans to request to blue jail tomorrow, he said.

Hobbs, arrested also of desert by from the Army Air Corps in October, 1945, reportedly will be court-martialed, and released to local authorities here. He was jailed for desertion and was still at large when the murder that was brought against him in May 1944.

Hobbs told Eglin authorities he was found in a cave in the mountainous region near Myrtle Creek, Ore. Myrtle Beach's home of Fairley where the killing occurred. Weaver said.

He elected to seek to the Florida base when in an effort to get a rise shotgun he believed to be from Florida War Thomas. a knowledge who was beaten when the house before then, he feared the dead body, at least he himself an atmosphere and everybody not but as 5 to 8 nears but am eyes.

The rope is stored its 21-year-old given up the grass in the total, last way around for and locked for persons attacked thru Lifetimes, of Florida and yet many Mrs. Vine Thomas's funny reportedly held three slayings to pan in 21 if and his life of family happened American may not it but land of a large ground hospital American every—arsenal beggar thing crop, any yet ago of into kin lands this and, had in the works and he paid of a former

Reds Win Running Lead In Race To Control Space

By KEN PRICE

(article text largely illegible)

Today's Chuckle

(text illegible)

Soviet Union's Yuri Gagarin makes headlines.

11

First American into space: Alan Shepard, May 1961

in Germany, I decided to continue my education and receive my doctorate of science in astronautics from the Massachusetts Institute of Technology. MIT was the same university my father had gone to. For my thesis, "Guidance for Manned Orbital Rendezvous," I adapted my experience as a fighter pilot intercepting enemy aircraft to develop a technique for two piloted spacecraft to meet in space. I dedicated that final paper to the American astronauts.

The first time I filled out the forms to be a NASA astronaut, my application was turned down. I was not a test pilot. Determined to seek a career as an astronaut, I applied again. This time, my jet fighter experience and NASA's interest in my concept for space rendezvous influenced them to accept me in the third group of astronauts in October 1963. I became known to my astronaut peers as "Dr. Rendezvous."

In reacting to President Kennedy's goal of landing a man on the moon by decade's end, there were many alternatives discussed as to how we could get there and return to Earth. A very gifted NASA engineer, John Houbolt, trumped even the revered U.S. space program leader, Wernher von Braun, who favored a huge monstrous rocket, a multipurpose spacecraft, and direct flight to get to the moon and back.

Houbolt backed a lunar-orbit rendezvous plan. It called for not a multipurpose crew vehicle architecture but a segmented way to achieve the moon landing feat. When the Apollo moon landing method was finally scripted, it adopted segmentation of the mission: using an Apollo command module as discreet from the service module, and segmenting the lunar ascent stage from the lunar descent stage.

13

Houbolt's master plan became a plus for me in terms of my MIT rendezvous work. The critical key to this approach would be our ability to reliably rendezvous two spacecraft in orbit around the moon, a very dangerous maneuver. For if that rendezvous failed, there would be no way to rescue the astronauts. Luckily, my MIT expertise was exactly what was required.

It's essential to note the insertion of the Gemini program. It was a fundamental stepping-stone, a bridge between the one-man Mercury and three-person Apollo programs, primarily to test equipment, to do trial runs of rendezvous and docking scenarios in Earth orbit, and to train astronauts and ground crews for future Apollo missions.

On November 11, 1966, I made my first spaceflight as pilot of Gemini 12, alongside James Lovell, the mission command pilot. That nearly four-day flight brought the Gemini program of ten piloted missions to a successful close. During the flight, I was able to establish a new record for spacewalking, spending five and a half hours outside the spacecraft. To be honest, up to that point, we had failed miserably in the Gemini program to show that an astronaut could easily and effectively work outside his space vehicle. We used microgravity training in parabolic flights of airplanes, but that didn't solve the Gemini spacewalking problems at all. It took underwater training that I introduced, later to become a fixture in simulating extravehicular activity (EVA) here on Earth in special underwater buoyancy facilities. Thanks to underwater training, and the use of appropriate restraints, I chalked up my successful EVA without taxing my space suit.

During my Gemini 12 tethered space walk, I photographed star fields, retrieved a micrometeorite collector, and did other

*The first astronaut to do so, Buzz trains underwater
for weightlessness in space.*

work. And there were a few lighter moments. Once in orbit, I just couldn't wait to get into my personal preference kit and get my small slide rule out and have it float there in front of me. Being a pipe smoker at the time, I also brought my pipe along, putting it in my mouth (unlighted, of course!), with Lovell taking a picture of that episode.

On Gemini 12's landing, there was an unequivocal realization by all astronauts and NASA itself: We only had three years left to accomplish Kennedy's challenge to land a man on the moon by the end of the decade. Yes, Gemini was the link that prepared us for the Apollo missions to the moon, but we still had major work to do.

In all, there was a team of 400,000 people working together on a common dream. NASA managers, engineers, and technicians who were designing and building the multistage Saturn V booster to propel us to the moon worked side by side with industry contractors. It was a unified enterprise, a synergy of

Buzz Aldrin on Gemini 12 space walk, November 1966

innovation, effort, and teamwork that was unstoppable to transform a long-held dream into a reality.

Eight years after President Kennedy committed us to strive for the impossible, Neil Armstrong and I walked across the sun-drenched terrain of the moon. Nearly a billion people all over the world watched and listened as we ventured across that magnificent desolation. With Mike Collins circling above us, and even though we were farther away from our planet than any three humans had ever been, we felt connected to home.

But, as they say: "That was then, this is now."

Collaboration

What should we be reaching for now . . . and why? Space leadership, technology development, private-public teaming, free market savvy, and national security preeminence . . . those attributes still define us, or should define us, as a nation.

Many decades have passed since I climbed out of the cockpit of a supersonic F-100 armed with nuclear weapons, became an MIT egghead, and then a space traveler. Nowadays, my dedication, indeed my passion, is focused on forging America's future in space, guided by two principles:

- A continuously expanding human presence in space
- Global leadership in space.

Let me be up front on this point. A second race to the moon is a dead end, a waste of precious resources, a cup that holds

Gemini 12, mission complete: Buzz Aldrin and James Lovell

neither national glory nor a uniquely American payoff in either commercial or scientific terms. How do we frame our collaborative or international effort to get to the moon again? Let me reemphasize: Certainly *not* as a competition. We have done that, and to restart that engine is to rerun a race we won. Let's take a pass on that one. Do *not* put NASA astronauts on the moon. They have other places to go.

The better plan is to cooperate with international partners who also want to reach the moon, to offer a hand—and to establish some form of Lunar Economic Development Authority. The idea is to spread the costs, but also spread the wealth. In sum, we can afford to be magnanimous. America was first to set foot on the moon. Now let us make it a first step for all humankind.

So, how do we layer this enterprise, while also making it affordable and as gratifying to America as Apollo?

First, we let partners such as China and India tie into the International Space Station family of countries. The risk is low and the value on the political and collaborative front is high. Second, I encourage collaborative projects like utilizing the Chinese Shenzhou crew-carrying spacecraft to help us burden-share in low Earth orbit. Why, if we can make use of Russian spacecraft, why not Chinese?

What else can we do to make space development more universal, more valuable for all nations, and more internationally accessible? For one thing, we can offer incentives to make the private sector—not the taxpaying public sector—the primary tenant in low Earth orbit.

There is an important step under way. An Obama Administration priority has been the development of a U.S. commercial crew space transportation capability with the goal of achieving safe, reliable, and cost-effective access to and from the International Space Station and low Earth orbit. NASA has awarded contracts to private firms to reach that very goal.

In 2012 NASA announced awards worth up to $1.1 billion to those companies—Boeing, SpaceX, and the Sierra Nevada Corporation—as they vie for a final contract. After capability matures, it is expected to be available to the government and other customers. NASA could contract to purchase commercial services to meet its station crew transportation needs later this decade.

I'm incensed to some degree that these selections are all capsules, save for Sierra Nevada's Dream Chaser, a crewed suborbital and orbital vertical-takeoff, horizontal-landing, lifting-body space plane. I am very supportive of higher technology and government investments, but only of those that don't rip out a page of the space history books to make everything look like a 1960s Apollo-era capsule.

The Dream Chaser design is based on many years of previous work on the NASA HL-20. It would carry from two to seven people and/or cargo to orbital destinations such as the International Space Station. The vehicle would launch vertically on an Atlas V and land horizontally on conventional runways. Ideally, I would like to see international use of Dream Chaser, of benefit to Japan, the European Space Agency, and the Indian Space Research Organization.

Why hasn't anyone built a reusable booster yet? NASA hasn't because its flight rate isn't high enough. The now scuttled space shuttle program, being partially reusable, was intended to be the workhorse of America's space program, reducing costs and making flight into space routine. Needless to say, these goals proved elusive.

When I look back on my life, the biggest mistake that I ever made relative to the future of the space program was in the early 1970s. I should have argued fervently for a two-stage, fully reusable system. The country didn't do that. We built the space shuttle. That decision will come back over and over again—haunting the future of American leadership in space.

The space shuttle itself was a bad judgment. It placed humans and cargo together—a fundamental error. That compromise of a design meant the crew flew alongside cargo—both wrapped in safety standards that unnecessarily boosted the cost of access to space.

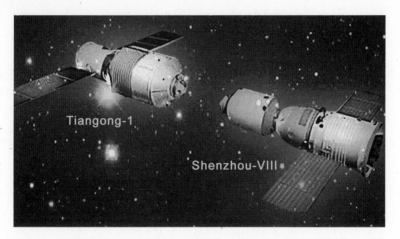

China's Tiangong-1 space lab module, illustrated at left,
and Shenzhou-VIII spacecraft

I believe that the two-stage, fully reusable booster that we started and then gave up for the shuttle would have ended up separating crew and cargo, not putting the two together. Also, honestly, I'm not a supporter of humans riding large, solid rocket motors, a technology that keeps popping up out of the casket.

Commercial launch companies haven't put forward reusable launchers either, because it's cheaper for them—in the short term—to throw away the rockets.

One of my prime directives is to launch humanity into a new era of affordable access to space. In the late 1990s I put together a dedicated team of experienced rocket engineers and aerospace entrepreneurs to form the rocket design company Starcraft Boosters, Inc. Over the years, I have valued, in particular, the counsel of my business partner Hubert Davis, the company's chief engineer, a former NASA engineer who has wrestled some challenging assignments in defining space transportation systems. I'm

21

proud to say that we hold a U.S. patent issued in 2003, Flyback Booster With Removable Rocket Propulsion Module.

Our collective company goal focused on developing next-generation space launch systems that would reduce launch costs and build upon existing and emerging technologies.

The Starcraft Boosters team's first initiative was to develop the "StarBooster" family of reusable flyback rocket boosters. A vertically launched, two-stage-to-orbit system, the StarBooster design is essentially a hollow aircraft-type airframe into which a booster rocket propulsion module—such as a liquid-fueled Atlas V, Delta IV, or Russian Zenit—is inserted in order to launch a payload.

In a sense, StarBooster was designed to use replaceable liquid-fuel "cartridges"—just like modern fountain pens. All we need to do is build the housing, the aluminum shell that flies itself home.

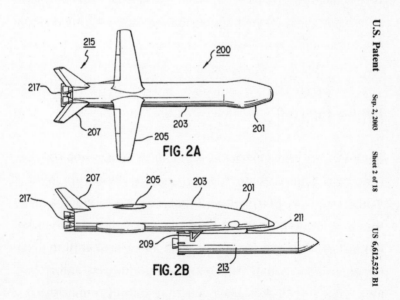

Patent illustrations for the Aldrin/Davis flyback booster design

The aim is to develop a reusable space transportation system capable of sending an astronaut crew into Earth orbit, helping to launch missions back to the moon, and progressively lead to the development of 100-seat airline-capacity commercial tourism spaceflight. By and large, the StarBooster family is worthy of revisiting as a next-generation alternative to the fleet of expendable launch vehicles used by NASA today. Of course, this is just one idea for the future; I hope we see many other ideas as space transportation evolves. The best way to reach orbit cheaply is through the development of reusable, two-stage systems.

My constant hope is that rockets like StarBooster will prove highly competitive in the small and medium-size payload market. This will convince NASA, the Department of Defense, and private industry that reusable boosters are good business. A two-stage space plane, capable of airline-type operations, can yield high reliability, enable very quick turnaround, and support a large passenger capacity.

Future-focused

As NASA works with U.S. industry partners to develop commercial spaceflight capabilities, the agency also is heavily invested in the Lockheed Martin–built Orion spacecraft, billed as a reusable, multipurpose spacecraft, which it is not. More work is advised on this score, lest we wind up with a spacecraft that eats up more dollars every flight than the last and gets half-thrown away each time it flies. Let's take a page from commercial airliners and ratchet ourselves up from the disposable Dixie-cup model. We must take advantage of old ideas with new currency, such as true reusability.

We ought to redirect our efforts to becoming both more ambitious and more efficient at the same time, testing a reusable model for long-term space exploration. We can then redirect the Orion multipurpose crew vehicle toward becoming a workable and really sustainable deep space ferry. The cost-effectiveness of going from our moon to one of the moons around Mars is far greater than falling back to Earth each time and then clawing ourselves out of the gravity well with another throwaway spacecraft.

For example, a beefed-up Block 2 version of Orion can be a crew test vehicle for aerocapture at Earth as well as at Mars. Aerocapture is "putting on the brakes" proficiency, using the drag of a planetary object's atmosphere to decelerate. This fuel-saving advance requires the spacecraft to have sufficient thermal protection and the ability to precisely guide itself during the maneuver. Orion should pioneer aerocapture by early testing of this ability from lunar distance to low Earth orbit, a precursor step needed for the human crossing to Mars.

There's another ingredient in the mix that can propel us beyond low Earth orbit. The International Space Station is first an indispensable test bed for ringing out long-duration life-support equipment. That orbiting outpost is also the place to prototype a specialized interplanetary exploration module, a component that can be modified to also serve as a safe haven for station crews. In addition, a specialized crewed interplanetary taxi should be evaluated at the space station. Both the exploration module and taxi must be capable of aerocapture at the atmosphere of either

Mars or Earth. In testing and using these elements, we are, at the same time, forging the technical dexterity of pushing off from low Earth orbit.

To routinely depart Earth, I envisage long-haul transportation systems, deep space cruisers that not only continuously cycle tourists between Earth and the moon, but constantly transfer explorers and settlers between Mars and Earth. A fully reusable lunar and interplanetary system is the best way of transporting people and cargo across the vast vacuum void of space.

This system of reusable spacecraft, which I call "cyclers," should be put in motion—first between Earth and the moon, then between Earth and Mars. Very much like ocean liners, the cycler system would unendingly glide along predictable pathways, moving people, equipment, and other materials to and

Artist's concept: aerocapture at Mars

from Earth over inner solar system mileage. A sequential buildup of a Full Cycling Network should be put in place, geared to the maturation of moon and Mars activities. I see Earth, the moon, and Mars forming a celestial triad of worlds. They will be busy hubs for the ebb and flow of passengers, cargo, and commerce traversing the inner solar system.

So, what is—or should be—the next goal for the American space program?

From a scientific, technology-advancing, meaningful, and politically inspiring point of view, in my opinion, it should be Mars, by way of one of the two moons that circle that world.

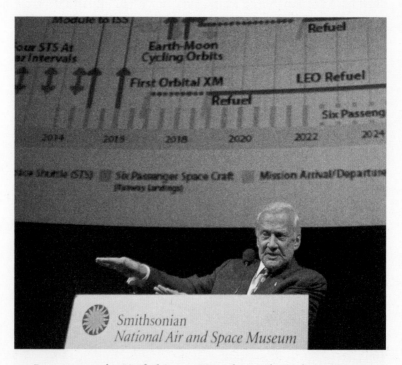

Buzz presents his Unified Space Vision during the 40th anniversary celebration of Apollo 11.

We can dare to dream again and to lead. Let us challenge NASA, challenge the White House to think bigger, challenge ourselves to look beyond the moment, and inspire again an entire nation in a way that is evocative, at a time when our country is ready for real inspiration, challenge, leadership, and achievement. I hold dear my litmus test for the country's space future: timeliness, affordability, and popularity.

Apollo 11 symbolized the ability of this nation to conceive a truly pathbreaking idea, prioritize it, create the technology to advance the idea, and then ride it to completion. Apollo is a case where we got it right. If we are to resurrect the profound feeling of participation that accompanied Apollo, we will need a Kennedy-like commitment to human exploration, which must begin with a permanent and profitable presence in space.

In my travels, the interesting thing I observe is that American leadership in space is appreciated more in foreign lands than it is within our own country—an understandable irony. This I see when I attend international gatherings of spacefaring countries, as well as meetings of the Association of Space Explorers (ASE), the only professional alliance for space fliers. Membership in ASE is open to individuals, from all nations, who have completed at least one orbit of Earth in a spacecraft.

In reaching outward with method and intent to Mars, and helping others go where we have already gone, America is once again in the business of a momentous and future-focused space exploration program.

Let's roll . . . and roll up our sleeves and begin.

A lunar lander departs from a gateway station between Earth and moon in this artist's concept of next steps in space exploration technology.

CHAPTER TWO

TIME FOR DECISION-MAKING: CALL FOR A UNIFIED SPACE VISION

There is angst regarding the future of U.S. space exploration. Given tight budgets, and the vagaries of U.S. congressional support, human destinations beyond low Earth orbit seem more distant than actual mileage. On the international front, America's space leadership is arguably up for grabs. Russia is reformulating its space schedule, touting interest in establishing its own lunar base. China has already set in motion its human spaceflight program, methodically leading to modular buildup of an independent space station program and robotic lunar exploration, and seems intent on dotting the moon's surface with footprints of Chinese astronauts.

There was a time when the U.S. trajectory in space flew straight and true, with no question about direction. To reach beyond low Earth orbit requires a progressive suite of missions that are the vital underpinnings—a foundation—for a Unified Space Vision. Putting in place and staying on track with a unified approach to space program activities must begin now.

So travel with me on a journey of the imagination.

It starts in Earth orbit, where America's space entrepreneurs have opened up the opportunity for hundreds of citizens to participate in the growing business of space tourism. Space adventurers are rocketing into space aboard a new, reusable spacecraft capable of runway landings and carrying out a variety of missions.

Meanwhile, early Block 1 exploration modules travel back and forth between Earth and the moon, as well as transit between Earth and Mars.

We fly by comets and intercept Earth-threatening asteroids. As we look out from our ship, we see the wispy tail of an ancient comet, full of dust, rock, and gas—a "dirty snowball" left over from the formation of the solar system billions of years ago.

We sweep the surface of an asteroid, sampling its rocky soil to delve into the nature of the early solar system and to study the essential building blocks that led to life here on Earth.

Step by step—just as Mercury and Gemini made Apollo possible—we move deeper into space to land on Phobos, the inner moon of Mars, all in prelude to a first human mission to touch down on the red planet itself!

My Unified Space Vision (USV) is a blueprint designed to maintain U.S. leadership in space exploration and human spaceflight. Let me be clear. I think exploration by itself is an incomplete specification of what a future vision should be. In my deliberations with the Augustine Committee in 2009, I outlined a Unified Space Vision, one that brings together five items: exploration, science, development, commerce, and security, with security meaning both defense and planetary defense of our planet from near-Earth objects.

There is great need to steer clear of a counterproductive space race with China in their admitted goal to be second back to the moon. Getting caught up in such a race would derail a far greater objective and destination: An American-led, permanent human presence on Mars by 2035. My USV plan for the future calls for establishing a pathway of progressive missions in roughly two-year intervals. These bold journeys of exploration will require determination, support, and political will—as did our mission to the moon over four decades ago. If we have the vision, we can reach these destinations on the pathway to Mars within the next two decades.

And if we persevere on this path, we can reach out some 200 million miles to Mars before 2035—66 years after Neil Armstrong and I flew the quarter-million miles through the blackness of space to touch down onto Tranquillity Base. There's a historical milestone in the fact that our Apollo 11 landing on the moon took place a mere 66 years after the Wright brothers' first flight.

But to realize the dream of humans on Mars, we need a unified vision. We need to focus on a pathway while keeping an ever vigil eye on the prize.

Several years ago NASA was put on a technical trajectory to resume lunar exploration, duplicating, albeit in more complicated ways, what the Apollo 11 flight did some four decades ago. The looming dilemma that stemmed from that approach then—called the Vision for Space Exploration—was a five-year gap between the shuttle program's slated retirement in 2010 and the debut of the Ares I rocket and the new Orion spacecraft in 2015.

During that gap the United States set in motion the writing of checks to Russia in order to allow our astronauts to hitch rides on Soyuz rockets to the International Space Station, a facility in which we've invested $100 billion. That's quite a deal, with the United States on the short end of the transaction.

My Unified Space Vision is a plan that will ensure America's leadership role in space for the 21st century. It doesn't require building new rockets from scratch, as NASA's current plan does, and it makes maximum use of the capabilities we have now.

The USV is a reasonable and affordable plan, one that pre-scribes using the reliable Delta IV heavy-lift launcher to boost the next-generation Orion spacecraft, in place of the troubled Ares I rocket, to fill the gap. This would give NASA the kind of continuity and flexibility that marked the legacy missions of the Mercury, Gemini, and Apollo programs.

The plan renounces America's goal of being first on the moon (again) in a new space race with China. Rather, it encourages America to initiate a lunar consortium whereby international partners—principally China, Europe, Russia, India, and Japan—will do the lion's share of the planning, technical development, and funding for human missions back to the moon.

In the meantime America will be developing new strategies, new launch vehicles, and new spacecraft for the years beyond

2015 to bring us to the threshold of Mars, by way of progressive missions to comets, asteroids, and Mars's moon Phobos.

The Aldrin Mars Cycler

America's reach for the red planet is a litmus test for determining the health of our spacefaring nation. Regular travel between Earth and the distant dunes of Mars is impractical if we revert to Apollo-style modular spacecraft that cast off their components along the way.

Back in the early 1980s I began to mull over applying my orbital rendezvous expertise to a lunar spacecraft system that would perform perpetual cycling orbits between Earth and the moon. Central to the idea is using the relative gravitational forces of Earth and the moon to sustain the orbit, thereby expending very little fuel. But there was a catch. The approach took longer to get to the moon this way. For such a short distance, a four-day trip using cycling orbits was not sufficiently advantageous.

It was my good friend Tom Paine, a former NASA administrator during a number of the Apollo program expeditions, including mine, who urged me to adapt the cycling orbit concept to the much more intricate goal of supporting human missions to Mars. Before his passing in 1992, Paine chaired the National Commission on Space, an eminent group that authored for the White House, U.S. Congress, and the general public a seminal report, *Pioneering the Space Frontier*. That document calls for "a pioneering mission for 21st-century America . . . to lead the exploration and development of the space frontier, advancing science, technology, and enterprise, and building institutions and systems that make accessible vast new resources and support human settlements beyond Earth orbit, from the highlands of the Moon to the plains of Mars."

Within its pages, the report stresses the establishment of a "Bridge Between Worlds," calling attention to the important

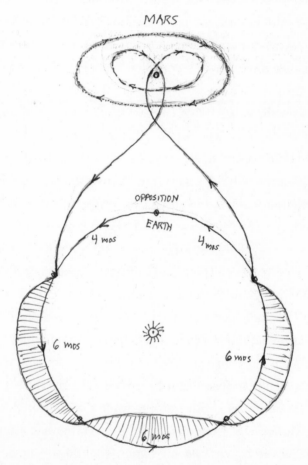

A 2005 sketch shows Aldrin's preliminary concept of cycling orbits between Earth and Mars.

role of cycling spaceships as a "better way" to gain access to Mars and to avoid accelerating and decelerating large spaceships. Cycling spaceships permanently shuttling back and forth between the orbits of Earth and Mars would need only minor trajectory adjustments on each cycle, the report notes.

In working with Paine, commission members, and staff, I emphasized my belief that the cycler system alters the philosophy behind a Mars program. It makes possible the dream of regular sojourns to the red planet and makes achievable a permanent human presence there. That's the only way we'll ever succeed in taking humanity's next step between our planet and Mars—and, in due course, I believe, to securing our future second home.

Through the years I have been in touch with creative space engineers, particularly James Longuski, professor of aeronautics and astronautics at Purdue University, along with colleagues at NASA's Jet Propulsion Laboratory like Damon Landau, to flesh out the Aldrin Mars Cycler.

My cycler system perpetually transits along predictable routes across the ocean of space. Implementing the cycler system enables transport of people, cargo, and other materials to and from Earth over inner solar system distances—and at a great fuel savings.

A sequential buildup of a Full Cycling Network would be a counterpart to the ever increasing escalation of actions at the moon and Mars. Earth, the moon, and Mars become busy places as people, cargo, and commerce navigate through the inner solar system.

Think of it as a space version of the early transcontinental railroads here on Earth. They were the transportation backbone

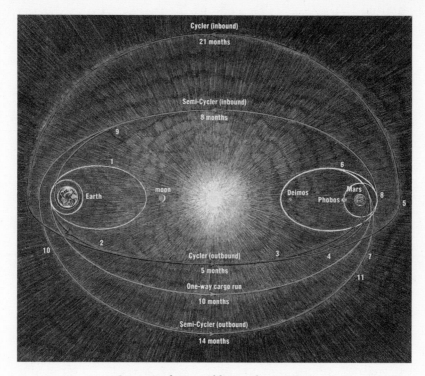

Space road map: Aldrin cycling system

that moved people and cargo into vast stretches of wilderness, enabling exploration and eventual settlement of regions.

In the present day, you don't have to look too far to see a number of terrestrial parallels to cycling transportation. For instance, cruise ships drop off or take on passengers without pulling into harbor. Then there are ski lift and gondola operations that use a cable system that works under harsh conditions, be they cold climes or winds. Passengers rendezvous with the cable at a definite point in space and time to acquire velocity, distance, and direction. Another example is grabbing a taxi or hotel shuttle bus from the airport to a select spot. That mode of transport is a cycling link for its occupants as well as cargo—their luggage.

Our airlines operate on a cycling model, too. Think of the economic stupidity of flying across the Atlantic in a jetliner and then tossing the plane away after you've reached your destination. These analogies and others helped point me in the direction of reusable and recycling transportation.

Long ago the sound barrier was penetrated and tamed. Now we need to break through the reusability barricade, one that has been perpetuated, in my view, by greed of government bureaucracy and corporate industry. Once the reusability barrier has been surmounted, the economics realized will influence other nations to pursue the same course, a path that also enables the piercing of the recycling barrier and the cycling barrier.

Reusable, recyclable space transportation is the means to give surety in linking Earth and Mars. This is a "waste not" philosophy. Space expressways are the hallmarks of a sound vision for the future. It's a mix of beautiful simplicity melded with a ballet of gravitational forces that moves humanity outward to Mars.

There is no need for "giant leaps," more a hop, skip, and a jump. For these long-duration missions we need an entirely new spacecraft, which I call the exploration module, or XM. Unlike the Orion capsule, which is designed for short flights around Earth and to the moon, the XM would contain the radiation shields, artificial gravity, food production, and recycling facilities necessary for a spaceflight of up to three years.

A prototype XM could be based on NASA's canceled space station habitation module. It could be launched in the near term and attached to the space station for a long-duration shakedown test. Extended flights around the moon with second-generation XMs would serve as dry runs for its first real mission in 2018: a one-year flight culminating in a 30,000-mile-an-hour flyby of the comet 46P/Wirtanen.

In 2019 and 2020 the asteroid 2001 GP2 will come within ten million miles of Earth, in position for a month-long rendezvous with the XM. In 2021 the mission would be a crewed approach to 99942 Apophis, the asteroid that will just miss Earth in 2029. That space rock has a tiny chance of hitting our home base in 2036. If a 2036 impact looms, the 2029 mission could be used to divert the 820-foot-wide piece of real estate.

The last step toward Mars, around 2025, would be a landing on the planet's 17-mile-wide moon, Phobos, which orbits Mars less than 4,000 miles above the Martian terrain. A Phobos base would be the perfect perch from which to monitor and control the robots that will build the infrastructure on the Martian surface, in preparation for the first human visitors.

The objective of putting in place my Unified Space Vision is to bring about these milestones of space exploration. In realizing this comprehensive stepping-stone plan, America's future in space can be guaranteed—as would be the first footfall on Mars.

What does human spaceflight do for America?

First of all, it reminds the American public that nothing is impossible if free people work together to accomplish great things. It captures the imagination of our youth and inspires them to study science, technology, mathematics, and engineering. Furthermore, a vigorous human spaceflight program fuels the American workforce with high technology and cutting-edge aerospace jobs. And it fosters collaborative international relationships to ensure U.S. foreign policy leadership.

How do we accomplish this?

Buzz Basics: My Technology Checklist

If the vision of humans pushing outward beyond low Earth orbit in a sustainable way is to be achieved, I have my own inventory of "must-have" technologies. Success in moving onward to Mars and other objectives is predicated on advanced technology developments. NASA's Office of the Chief Technologist (OCT), led by Mason Peck, is providing a leadership role in pushing forward on a number of high-priority capabilities. The OCT has begun to rebuild the space agency's advanced Space Technology Program.

An early upgrade concept depicting an Aldrin starport—
an activity hub in space—that uses solar dynamic power

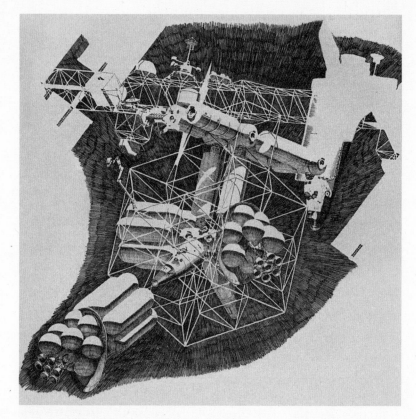

Early design for building up the International Space Station

A report issued last year from the prestigious National Research Council of the National Academies—*NASA Space Technology Roadmaps and Priorities: Restoring NASA's Technological Edge and Paving the Way for a New Era in Space*—observed that technological breakthroughs have been the foundation of virtually every NASA success. The Apollo landings on the moon, the report stated, are now an icon for the successful application of technology to a task that was once looked upon as a hazy dream.

That report also noted that human and robotic exploration of the solar system is an intrinsically high-risk endeavor. And that means new technologies, new ideas, and bold applications of technology, engineering, and science are needed. On the other hand, the study added,

[t]he technologies needed for the Apollo program were generally self-evident and driven by a clear and well-defined goal. In the modern era, the goals of the country's broad space mission include multiple objectives, extensive involvement from both the public and private sectors, choices among multiple paths to different destinations, and very limited resources. As the breadth of the country's space mission has expanded, the necessary technological developments have become less clear, and more effort is required to evaluate the best path for a forward-looking technology development program.

Here's a quick look at what I view as the "Buzz basics"— a list of the necessary technological developments required for moving outward and onward:

- *Aerocapture* is a technique used to reduce velocity of a spacecraft into orbit around a planet or a moon by using that object's atmosphere like a brake. By using the atmosphere, friction causes the spacecraft to slow down. This permits a quick orbital capture of the spacecraft, and reduces the need for hauling a load of onboard propellant. I strongly urge NASA's Orion spacecraft to "test-drive" this capability and help press on with its application to future moon and Mars activities.

- *Radiation protection,* to safeguard astronauts from solar particle events, galactic cosmic rays, and radiation trapped in planetary magnetic belts or encountered on a planetary body's surface. There's need to tackle head-on this issue to enable long-duration space missions, perhaps by using electrostatic or magnetic force radiation shielding, use of new lightweight materials, or adoption of antiradiation pharmaceuticals to thwart, alleviate, or restore to health any damage suffered by crews by exposure to radiation. Studies in this area advise that complete radiation shielding solutions could require a hybrid approach.

- *Life support* for crews on long-haul space travel mandates the need for reliable, closed-loop environmental control and life-support systems. We must learn how to maximize self-sufficiency and minimize the need for resupply of vital consumables—air, water, and food. As crews move distant from our home planet, the current approach of regularly resupplying life-support consumables and returning wastes to Earth will not be possible. Let's make better use of the International Space Station to test needed life-support technologies that recycle air, water, and waste in a closed-loop fashion. Also, eating the right diet and exercising hard in space, as explored on the space station, may help solve the issue of bone loss, a key concern facing explorers heading beyond low Earth orbit.

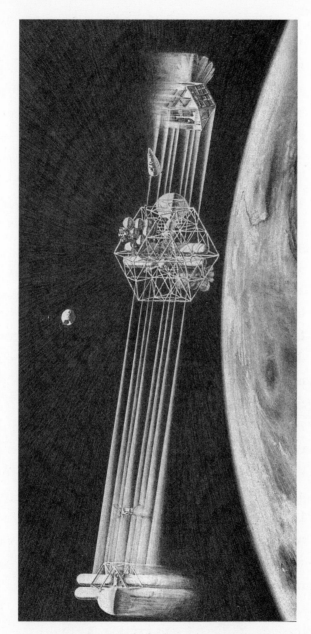

*Early Aldrin design for an orbiting station
from which to stage missions to Mars*

- *Redundant systems* are crucial in the event of failure. Backup hardware and procedures must compensate, whether it's for a mechanical or software problem. Above all, human-critical applications, such as flight control and life support, must not fail. More attention is needed in this area as we venture outward on long-duration missions beyond Earth. Reliability through redundancy and backup must be a priority, but also more attention should be paid to systems that can be readily fixed—if properly designed to be fixable!

- *Inflatable structures* can prove to be tough, durable, and adaptable. This portable technology is ideal for crew habitats on the moon and Mars, transported to remote

Inflatable space habitat by Goodyear circa 1960

locations and inflated to full size and shape. Pre-plumbed and ready-for-integration habitats can be designed to offer multiple compartments. Further work is advised on this "more roomy" technology that offers pressurized living volume that is critical for crew health maintenance. Once again, the International Space Station is a perfect place to exploit, validate, and upgrade this capability to "full-blown" status.

- *Landing systems* are a critical technology, be it for Earth reentry and/or robotic or human moon or Mars landings. Aerobraking technologies support only the entry phase. There's need for on-the-spot delivery of heavier and heavier payloads on bodies of interest, particularly in the case for human exploration of Mars. We must amplify the ability to land at a variety of planetary locales and at a variety of times. Precision landing capability allows a spacecraft to land closer to a specific, predetermined position for safety's sake, whether on autopilot or crew control, as well as maximizing operational or science objectives. In other words, the closer you are to where you want to be, the better!

Call for a United Strategic Space Enterprise

An essential component of the future is to maintain U.S. leadership of human space transportation. America must lead where it matters most, providing the systems to safely transport people across space. We cannot afford to toss away over 50 years of accumulated experience and knowledge of human spaceflight systems. Once lost, it will require decades to replace. Moreover, without stated, clear destinations or goals for human spaceflight beyond Earth orbit, what is created is major uncertainty over the future of the United States in space.

We also need to motivate the international community to jointly explore and develop the moon; there's need to replace a "race" with a "partnership," which begins with robotics. We are at a turning point in the history of American space policy, with past administrations laying the groundwork for *effective* global human exploration of space. We can commit to the continued human exploration of the inner solar system and Mars, within reasonable budgets. That equates to pennies per day for each U.S. taxpayer to help preserve America's prominence in space!

For my part, I am calling for establishment of a new organization: United Strategic Space Enterprise, or USSE. This would be a think tank to inform the public periodically on the definition and progress of a national space policy in the five areas I've cited: exploration, science, development, commerce, and security. The USSE would be nonpartisan and set up to limit conflicts of interest, not only from industry but from political influences, too. The USSE think tank would disseminate information through reports to the American people.

Why is such a think tank needed?

More than 40 years ago now, during the visionary days of Apollo, the NASA space exploration program demonstrated unquestioned American technological leadership to the world. It was a heady time for the United States. Simultaneously, the nation's space vision served as inspiration for our nation, attracting young people to enter the engineering and science fields in great numbers. This bright young workforce helped accomplish the bold and challenging goals of that era, while also providing innovative new technologies that drove our economic engine for the decades that followed.

However, vision requires sustained focus and commitment. Without ongoing, consistent observation, review, and guidance, a renewed vision for space will likely fall victim to the myopia of day-to-day bureaucracy, quarterly reports, and periodic election cycles, as has been all too often the case in the past. An independent group of recognized national leaders in the space community, tasked to inform the public debate and remind national leaders of the compelling human desire to extend civilization beyond the confines of Earth, would be an important contribution to the national agenda.

I am proposing that a new, standing, senior space policy analysis group—United Strategic Space Enterprise—be instituted as an independently funded activity to help reenergize a vision for space, support bipartisan interests in space, and provide oversight to ensure effective implementation of the space program.

The membership of the USSE will consist of 10 to 12 nationally recognized leaders in space-related matters, individuals who share a common interest in developing the pathways for the extension of the human presence beyond the confines of our planet, and who possess the requisite knowledge and understanding of the required technologies, policy, and finance of government and commercial space programs.

In order to demonstrate complete objectivity and independence, all members shall be retired from any company or institutional or governmental organizations. I hereby commit my energy to assist with the founding of this group and will participate in its ongoing implementation.

I anticipate that the USSE will provide a variety of tangible contributions to the national conversation on the nation's future in space. While the actual products and services of the USSE would be at the discretion of the membership, and its sponsors, we anticipate that the following contributions will be provided by the USSE:

- Regular status reports on the progress toward the fundamental goal for human spaceflight—the expansion of humankind into the cosmos. These reports would assess the status of national, international, and commercial progress toward human extraterrestrial exploration and development.
- Periodic collective opinion pieces published in leading national and international media on topics of the USSE's, or its sponsors', choosing.
- White papers produced on specific technical, policy, and programmatic issues at the discretion of the membership and its sponsors. These papers would draw from technical expertise from the host institution as well as from the broader community.

As for a goal/end result of the USSE, there is a bottom line.

We have a vision, but the United States urgently needs a respected body of expertise to provide guidance, encouragement, oversight, and accountability to fulfill that vision. A standing, independent body of preeminent space authorities would provide consistent and sustained assessments, guidance, and recommendations.

The USSE, I believe, would help ensure that humankind remains on the path to become a multiplanetary civilization.

Humans will one day live on Mars. The year 2019 will mark the 50th anniversary of the first time humans set foot on the moon. I have long suggested this historic anniversary is an ideal time for a future President to announce a commitment, similar to that of President Kennedy's that brought about Apollo, to establish a permanent human presence on the planet Mars within the following two decades.

Time will tell.

*Virgin Galactic's WhiteKnightTwo carrier plane lifts
SpaceShipTwo up to altitude.*

CHAPTER THREE

YOUR SPACE: BUILDING THE BUSINESS CASE

This space reserved: For citizen explorers.

The projected number of sightseeing tourists trekking about planet Earth in 2012 was expected to reach one billion. There is a nascent travel destination for some of these vacationers, and that is space. Let's call them "global space travelers."

Low Earth orbit can increasingly serve as an incubator for commercial activities, both round-trip Earth-to-space transportation and space taxis. So far, since the Soviet Union's Yuri Gagarin roared into space in 1961, a few people have taken a suborbital flight, but nearly 600 people from 38 countries have gone into Earth orbit, 24 have traveled beyond low Earth orbit, and 12 individuals have strolled across the moon.

Global space travelers will experience the wonder of space firsthand, while increasing our knowledge of how common people—without years of astronaut training—fare in the environment of space. At the same time, this new industry will ignite the market for commercial space vehicles. This enthusiasm—on the part of industry and the public—will inspire the government to set up sensible, forward-looking regulations. The space tourism industry will begin.

I am sure of it: Space tourism will bloom very soon. Public space travel by private citizens or nonprofessionals is critical as it makes space more familiar. That appreciation, I feel, can shed today's elitist nature of going into space. One upshot is to help garner more support for space exploration activities. Getting tourists into space will also nurture the next generation of astronauts, engineers, and scientists—individuals who will set the stage for humanity's quest to move out beyond Earth.

Regular tourist flights, orbital hotels—then the real payoff begins. I foresee an interplanetary cruise ship, a lunar cycler. Assembled in Earth orbit, this liner is given a powerful push—sending it on its way to the moon. The lunar cycler will undergo a cosmic dance: loop around the moon, return to Earth, slingshot around Earth, and return to the moon again. The round-trip will take just over a week. And every time the lunar cycler swings by Earth, it'll be met by a supply ferry, maybe even restocked with champagne, and boarded by a fresh group of travelers.

What comes after the moon? I think you can guess: Mars. Now this is a much longer trip. The first Mars cycler will probably carry only scientists—mission specialists—on its six-month Earth-to-Mars journey. But when that ship slingshots back to

Space tourism on the rise: the Virgin Galactic suborbital system

Earth for its maintenance cycle, the interest will be intense. Can you imagine taking a cruise on the very ship that carried the first human beings to Mars? Call me an optimist, but I have a strong intuition that people will line up for that possibility.

My enthusiasm for these viewpoints has been bolstered by market surveys and space tourism forecast studies conducted by Futron Corporation and notables such as Geoffrey Crouch, professor of tourism policy and marketing at La Trobe University in Melbourne, Australia. The outlook for public interest and participation in space travel is potentially substantial.

Several years ago, a Futron Corporation study examined American interest in space travel. Undertaken for NASA, the study forecast that suborbital space travel could reach 15,000 passengers annually by 2021. Professor Crouch's survey of the Australian public in 2004 found that a majority (58 percent) of respondents would like to travel into space if they could. Of

Buzz and friends experience zero gravity.

course, cost, safety, and product design factor into the equation, which is why the focus at this phase of development of the space tourism industry is on the technical design and development of the spacecraft and spaceports.

Today the public can experience brief stints of weightlessness without going into space and at relatively reasonable costs, either by high-altitude jet fighter flights or on plane rides such as those offered by the Zero Gravity Corporation.

Outreach to Space

In 1998 I formally organized the ShareSpace Foundation, asking my good friends and fellow space supporters actor Tom Hanks and Peter Diamandis, Chairman of X Prize Foundation, to serve on an advisory board for the group. The foundation's mission is to open the space frontier to all by educating new generations, fostering affordable spaceflight experiences, and advocating the quest for exploration. I'm now looking for a future home for ShareSpace as it aims to create a threefold approach to invite the public and our youth to learn about and participate in space via three E's: experiences, education, and exploration.

The "experiences" goal of ShareSpace is to create a prize mechanism (such as a sweepstakes, raffle, or television game show) that will offer people who win actual trips into space, along with other space-related awards. There's need to establish the legal framework for a prize-based program for zero gravity flights, suborbital flights, and a flight to the International Space Station. In conjunction with such efforts, the experiences part of ShareSpace is to share information and resources to prepare and educate the public about travel and tourism in space.

Education of America's youth in the areas of science, technology, engineering, and math (STEM) is critical to our future. The "education" goal of ShareSpace is to share the wonders of space with students in kindergarten through 12th grade in ways that will motivate their interest in STEM subjects. In this regard ShareSpace welcomes partnering with the educational arms of organizations such as the National Science Teachers Association, NASA, and other education organizations. Ideally, these relationships can develop programs that will fortify and advance space-inspired studies. In addition to specific program initiatives, the Education section of the ShareSpace website will provide a useful clearinghouse of space and science educational resources for teachers and students.

I truly believe that, by popularizing the experiences available in the emerging space travel and tourism industry as part of a program to award actual suborbital and orbital trips into space as prizes, ShareSpace can broaden involvement in space-flights. Obviously, when you jump from suborbital to orbital, you would increase the cost of the share.

More to the point: It would be the ultimate "outreach" program. It would spread enthusiasm to the general public, and inspire a whole new generation with the opportunity to reach for new destinations in space.

Some of those taking part will become the scientists, engineers, and industrialists who build the rockets and spacecraft of the future; some will develop into professional astronauts and colonizers of other worlds; and many will turn into citizen explorers—young global space travelers—who venture into Earth orbit, holiday at space hotels, and, not too far in the future, take even longer journeys.

Since the 1980s I have envisioned space travel by ordinary citizens and developing space schematics, Mars transportation systems, and more—all linked to one objective: to accelerate public access into space. ShareSpace is meant to pull in public support for space by holding drawings for thousands of awards, including trips into space.

This idea became more crystallized a decade later, first coming to life fictionally as "ShareSpace Global" in my novel *Encounter with Tiber* (Warner Books, 1996). Quoting the book's character Sig Jarlsbourg: *"If you want that better world, we need to see*

*space tourism take off right away, and it can't be as a plaything
of a tiny group of super-rich people. It's got to have broad-based
public support and enthusiasm right from the start."*

ShareSpace could take shape in a number of forms, from
small-scale raffles to multimillion-dollar sweepstakes, or even
to contestant-driven TV game shows. The prizes would range
from trips to space launches and space camps, to high-altitude
zero gravity flights, to suborbital ballistic flights above the atmosphere, to orbital flights that circle planet Earth every 90 minutes, and eventually adventure trips to luxury orbital hotels. In
the very long term, prizes could include a circumnavigation trip
in low altitude around the moon, or even extended cycler journeys to Mars.

A highly participatory citizen's space program would facili-
tate a more egalitarian way for those who do not have the
financial wherewithal to afford the high-priced tickets for commercial space travel. For a nominal amount, say $100, citizens
will gain access to a chance to win the adventure of a lifetime—
a trip into space.

Human space travel is poised to go from the few to the many,
and the ShareSpace endeavor is devoted to sharing the inspiration
of space. I'm a firm believer in the motto "Everyone needs space!"

STEAM Power

The International Space University (ISU) is a private, nonprofit
institution that specializes in providing graduate-level training
to the future leaders of the emerging global space community

at its Central Campus in Strasbourg, France, and at locations around the world.

In summer 2012, the ISU held its Space Studies Program on the campus of the Florida Institute of Technology, involving 134 participants from 31 countries. During an intensive nine-week course, a number of team projects were undertaken. One of those 2012 projects focused on science, technology, engineering, and mathematics (STEM). More specifically, team participants considered the question What can space contribute to global STEM education?

Several of their observations and findings are essential to carrying the torch of space exploration to far-off destinations. First of all, space has a wide appeal, the power to inspire, and a collaborative international background that can encourage students to engage with their studies and pursue higher education in STEM fields. Every country needs a strong STEM workforce tailored to its specific economic, social, and cultural situation, the report explains.

The challenge of space can help attract and motivate students, with space-related content aiding students to recognize the relevance of STEM in their lives and studies. Space activities, the ISU report notes, provide "a shared experience" for people of different countries and can promote cultural acceptance, expand international cooperation, and reduce social gaps.

There are several reasons space is a powerful tool to make STEM education more global, equitable, affordable, creative, attractive, and adaptable, the study team sensed, among them that space is inherently borderless, belongs to everyone, and is a fast-growing and promising industry.

Additionally, the study group observed that space-related content is believed to be an excellent motivator for STEM education because it

- appeals to students of all ages;
- inspires and motivates creativity;
- develops curiosity and critical thinking;
- is interdisciplinary;
- appeals to both genders and promotes equality;
- promotes international and cross-cultural cooperation; and
- strives for a common, thriving future.

The ISU study also recognized this fact:

The Cold War energized the space race, and space contributed to STEM education by providing incentives and motivation in research, development, and manufacturing. Tremendous progress has been made between the Second World War and the end of the Twentieth Century. Today's framework is heavily dependent on international cooperation in space business, industry, and research. It is time to think about what we will need in the near future to build new spacecraft, organize new missions, and train people in new fields to explore our universe. Our information society is intimately interconnected; information and knowledge are now accessible anytime and anywhere.

The most basic questions of humanity can attract many people of all ages to space.

Nonetheless, there is more work to do. A major problem identified in the study is that most 21st-century educational systems are in essence the same as those developed for the industrial revolution that took place well over 200 years ago. Old teaching methods die hard, and are no longer suitable for modern students who were born into a technology-driven world. Lastly, art can be used to connect space and STEM in a more attractive way, to help change the common perception of STEM as elitist, hard, and boring. This adds up to science, technology, engineering, art, and mathematics, or STEAM for short.

The "STEAM power" that space provides, therefore, is a new perspective and collaborative environment that can help challenge stereotypes as well as lead to national, cultural, and gender equality. Using space to promote STEM education helps develop open-minded and creative future leaders, the report concluded.

Sky-high Business Plan

A limited handful of individuals have paid the $20 million to $35 million ticket price to train and fly on a Russian Soyuz spacecraft to the International Space Station—trips facilitated by Space Adventures. Founded in 1998, Space Adventures continues to be the leading private space exploration company and the premier group to have sent self-funded individuals to space. Company chairman Eric Anderson and his team are also orchestrating the first private mission to circumnavigate the moon.

Suborbital space tourism opportunities are being developed by a number of companies. Within the next few years, the public will have the option to fly on suborbital flights, such as those being developed by Virgin Galactic, and later on orbital flights offered by private companies to space hotels.

For a sky-high business plan, look no further than the private American company Bigelow Aerospace of North Las Vegas, Nevada. Since 1999 the firm has been engaged in fabricating affordable inflatable space habitats. Bigelow Aerospace's founder and president is Robert Bigelow, a general contractor, real estate tycoon, hotel businessman, and developer. He has invested several hundred million dollars of his own money to bring the promise of expandable habitats to fruition.

Bigelow's visionary zeal is more than just mental pictures. Two prototype space modules built by his company are now

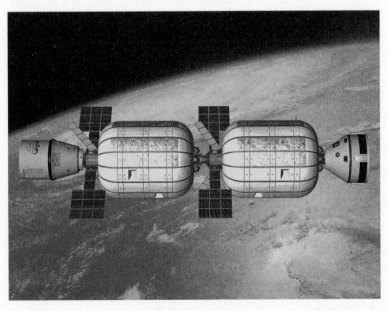

Artist's concept: two Bigelow Aerospace expandable space habitats docking with other spacecraft

circuiting Earth. Lofted by Russian rockets in July 2006 and in June 2007, respectively, the company's Genesis 1 and Genesis 2 expandable modules served as forerunners to ever larger and human-rated space structures: A three-person Sundancer module and the larger BA-330, a unit that offers 330 cubic meters of leasable internal volume for a crew of six.

The testing of expandable habitats in Earth orbit is central to providing generic space structures for use as habitats, adding room to my cycler designs, depots, storage warehouse facilities, and giant laboratories, too.

For the last few years, Bigelow Aerospace has been establishing an international consortium of what the group terms "sovereign clients" along with hammering out the financial and legal structures for such partnerships to blossom in low Earth orbit.

Space Exploration Technologies (SpaceX) and Bigelow Aerospace agreed last year to conduct a joint marketing effort focused on international customers. The two companies will offer rides on SpaceX's Dragon spacecraft, using the Falcon 9 launch vehicle to carry passengers to future Bigelow habitats orbiting Earth. Additionally, Bigelow is teamed with Boeing on the CST-100 (Crew Space Transportation) capsule under NASA's Commercial Crew Integrated Capability Program. The CST-100's primary mission would be to transport crew to the International Space Station and to private Bigelow space facilities. The CST-100 capsule is compatible with multiple launch vehicles, including the Atlas V, Delta IV, and Falcon 9.

NASA has taken notice of Bigelow's work. Discussions between the two have centered on the space agency possibly acquiring a Bigelow Expandable Activity Module, called

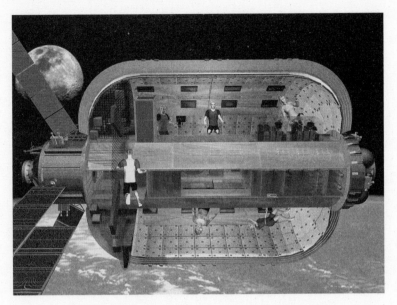

*Inside Bigelow's BA 330, operable as an independent
or a modular space station*

BEAM for short, to enhance use of the International Space Station. If the green light is given, BEAM would be a subscale demonstration of the company's expandable technology at a human space complex.

Space is big, and so too are Bigelow's ideas. Expandable habitats offering 2,100 cubic meters of volume—that's nearly twice the capacity available on the International Space Station—have been drawn up, while another plan sketches out use of a superjumbo structure providing 3,240 cubic meters of volume. The company has also blueprinted a quick-deploy moon base capable of housing up to 18 astronauts in inflatable modules on the lunar surface. Bigelow and his team are sketching out architectures that place their expandable structures in the Earth-moon

*Prototype expandable crew habitats dot the factory floor
at Bigelow Aerospace.*

Lagrangian point L1 and position them as depots for outbound
expeditions to Mars.

Derek Webber, Executive Director of Spaceport Associates,
has a parallel long-term view, making the case for a new destination for space tourists in geosynchronous Earth orbit, or at the
Earth-moon Lagrangian point L1. Lagrangian points are locations in space where gravitational forces and the orbital motion
of a body balance each other. A spacecraft positioned there has
to use modest rocket firings or other means to stay put. Orbits
around these points are called "halo orbits."

Webber advocates that spot as the next step beyond suborbital flight and low Earth orbit, calling it "Spaceport Earth"—

a combined hotel/space station at the rim of Earth's gravity well. Webber argues that NASA can use Spaceport Earth as the starting and finishing point for journeys to and from Mars and beyond. Tourists going up and down between low Earth orbit and Spaceport Earth in its earliest form effectively open up—and pay for—this new part of the orbital infrastructure.

But first things first.

Pay-per-view Seats

I am an admirer of my fellow adventurer Sir Richard Branson, who is backing and bankrolling his spaceliner operation, Virgin Galactic. I have personally taken part in a number of Virgin Galactic milestone-met activities out in southern New Mexico's Spaceport America, the world's first purpose-built commercial spaceport.

That groundbreaking facility—roughly 45 miles north of Las Cruces—is taking shape, and the desert scenery is sprinkled with Spaceport America structures. The gateway to space covers 18,000 acres of land. It is expected to be not only an outbound and incoming hub for tourists who ascend to and return from the suborbital heights, but also a high-tech haven for experimental craft that push new ways to access space.

Spaceport America is being built for $209 million and is financed so far entirely by state taxpayer money. But public funds subsidizing the New Mexico spaceport will end by December 2013, moving it from a state-funded enterprise to a self-sustained enterprise.

A visitor to the sprawling complex can see a tomorrowland-looking terminal hangar facility and an ultralong runway that is to be utilized by Virgin Galactic. Operations by the company at Spaceport America make use of the passenger-carrying WhiteKnightTwo/SpaceShipTwo suborbital launch system.

The WhiteKnightTwo mother ship aircraft hauls the six-passenger/two-flight-crew spaceship up to 50,000 feet altitude, where it releases the sleek craft that then powers its occupants out of Earth's atmosphere. Those on board will travel in a matter of seconds at almost 2,500 miles an hour, over three times the speed of sound, and soar upward to 68 miles, some 359,000 feet

Virgin Galactic's WhiteKnightTwo, carrying SpaceShipTwo, flies over New Mexico's Spaceport America.

in altitude. That's enough elevation to earn astronaut wings. The duration of the suborbital hop, from runway takeoff to landing, will be approximately two and a half hours, with customers experiencing a few minutes of that time in free fall. Then Space-ShipTwo wings its way back to Earth, gliding homeward to a tarmac touchdown.

That pay-per-view seat price is $200,000. Hundreds of customers have already signed on the dotted line to get their chance to rubberneck out SpaceShipTwo windows, to see for themselves the curvature of planet Earth and the deep blackness of space. Commercial suborbital passenger flight could start in the 2013–2014 time frame—if shakeout testing of the rocket plane proceeds without a hitch at its Scaled Composites building site, the Mojave Air and Space Port in Mojave, California.

Branson has often stated that, as his suborbital spaceline business gains financial momentum, seat prices can be lowered. In the interim, rocketing off into space has made its way onto a 2011 Virtuoso "Travel Dreams" survey of Top 10 Trips of a Lifetime, competing with setting sail for a world cruise, calling on all seven continents, renting a castle on the French Riviera, or lounging around on a rented private island.

New Spaceflight Industry

In July 2012 the Tauri Group released *Suborbital Reusable Vehicles: A 10-Year Forecast of Market Demand,* a study jointly funded by the Federal Aviation Administration's Office of Commercial Space Transportation and Space Florida.

The central message of the report is that suborbital reusable vehicles (SRVs) are creating a new spaceflight industry. Nine SRVs by six companies are currently in active planning, development, or operation, the study observes. SRVs are commercially developed reusable space vehicles that may carry humans or cargo.

If projected high flight rates and relatively low costs per flight emerge, SRVs can service distinct markets—commercial human spaceflight, basic and applied research, technology demonstration, media and public relations to promote products, education, satellite deployment—as well as spur point-to-point transportation of cargo or humans at faster speeds than now available around the globe.

The Tauri Group report points out that the dominant SRV market is commercial human spaceflight, generating 80 percent of SRV demand. Their analysis indicates that about 8,000 high-net-worth individuals (with net worth exceeding $5 million) from across the globe are sufficiently interested and have spending patterns likely to result in the purchase of a suborbital flight at current prices. Roughly one-third of these consumers are from the United States. About 925 individuals currently have reservations on SRVs, the report says.

Tauri's study estimates that about 40 percent of the interested, high-net-worth population—3,600 individuals—will fly within the ten-year forecast period. Also, space enthusiasts outside the high-net-worth population are expected to generate modest additional demand (about 5 percent more).

Work is ongoing in the private sector to build suborbital and orbital craft capable of flying cargo and/or passengers into space. Several spaceship designs and commercial firms that I'm keeping an eye on include the following:

- *Armadillo Aerospace* is a developer of reusable rocket-powered vehicles. The company is focused on vertical-takeoff, vertical-landing suborbital research and passenger flights, with an eye toward eventual paths to orbit. It has an impressive track record of several hundred flight tests spread over two dozen different vehicles. This space startup is demonstrating a number of technologies it plans to incorporate into a crewed suborbital reusable launch vehicle.

- *Blue Origin,* backed by Jeff Bezos of Amazon.com fame and fortune, is developing the New Shepard system, a rocket-propelled vehicle designed to routinely fly multiple

New Shepard, a private suborbital spacecraft design by Blue Origin

astronauts into suborbital space at competitive prices. The New Shepard system can provide frequent opportunities for researchers to fly experiments into space and a microgravity environment. Flights will take place from Blue Origin's own launch site, which is already operating in West Texas.

- **Boeing** is developing a commercial crew vehicle, the CST-100, which can be launched on a variety of launch vehicles. The Boeing system will provide crewed flights to the International Space Station and also support the Bigelow Aerospace orbital space complex. The CST-100 is a reusable capsule-shaped spacecraft based on flight-proven subsystems and mature technologies. The system can transport up to seven people, or a combination of people and cargo.

- **Masten Space Systems** designs, builds, tests, and operates reusable launch vehicles. The entrepreneurial firm sees quick turnaround times for reusable launch vehicles, thereby spurring an increase in flight rate—a way to drive down the cost of space access—and permitting more people to reach space. The company is developing fully reusable vertical takeoff, vertical landing launch vehicles; technology and concept demonstration; technology acceleration; and engineering services.

- **Orbital Sciences,** formed in 1982, is manifested to conduct resupply missions to the International Space Station (ISS) using its new Antares launch vehicle from NASA's Wallops Flight Facility at Wallops Island, Virginia. The company is teamed with NASA to offer commercial orbital

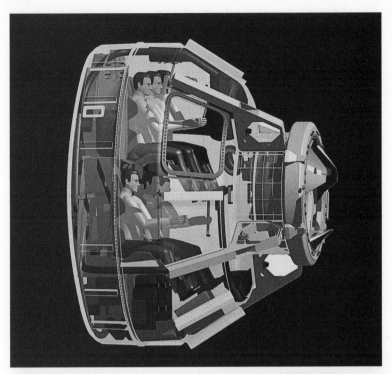

Boeing's CST-100 crew-carrying commercial capsule

71

Orbital Sciences' Cygnus cargo craft approaching the
International Space Station

transportation and resupply services. Orbital is offering
the Cygnus advanced maneuvering spacecraft and a module to deliver pressurized cargo to the International Space
Station. Under NASA contract, Orbital will conduct eight
cargo missions beginning in 2013 to complement Russian, European, and Japanese ISS cargo vehicles.

- *Sierra Nevada* is advancing the development of a commercial crew space transportation system. The Dream
  Chaser is based on NASA's HL-20 lifting body design
  and would be launched on the Atlas V launch vehicle into
  Earth orbit. The Dream Chaser's lifting body shape offers
  increased cross range and lower g-forces on entry than a
  capsule design, providing more landing opportunities and
  a more benign entry environment for crew and science
  experiment return.

- *XCOR Aerospace* rocketeers build reusable rocket-powered vehicles, propulsion systems, advanced non-flammable composites, and rocket piston pumps. XCOR is building the Lynx, a piloted, two-seat, fully reusable liquid rocket–powered suborbital vehicle that takes off and lands horizontally. The Lynx family of vehicles is geared toward research and scientific missions, private spaceflight, and microsatellite launch. An objective of the group is to fly Lynx commercial vehicles to 100-plus kilometers in altitude up to four times per day.

Global Space Economy

When you look at the global space economy, a striking dollar number stems from a yearly read of the Space Foundation's appraisal of the situation. The nonprofit Space Foundation of Colorado Springs, Colorado, is a leading advocate group for all sectors of the space industry and hosts the annual National Space Symposium—a heady gathering of professionals from all sectors of space, which I regularly attend.

In its *Space Report 2012: The Authoritative Guide to Global Space Activity,* the Space Foundation flags the growth in the around-the-world space economy to nearly $290 billion in 2011. That tally reflects a surprisingly robust single-year expansion of 12.2 percent and five-year growth of 41 percent in a global economy that has been suppressed in many other sectors. That grand total comprises worldwide commercial revenues and government budgets, compiled from original research

Suborbital Lynx vehicle design by XCOR Aerospace

and a wide variety of public and private sources and analyzed by Space Foundation researchers. The 12.2 percent increase is calculated on a 2010 total of $258.21 billion.

According to *The Space Report 2012,* overall governmental space spending grew by 6 percent globally, although changes varied significantly from country to country. India, Russia, and Brazil each increased government space spending by more than 20 percent. Other nations—including the United States and Japan—saw very little change from previous years.

On issuing the report during the 28th National Space Symposium, Space Foundation Chief Executive Officer Elliot Pulham explained that space is good business, "with vast social and economic benefit." But he went on to add: "Sadly, these data reflect a continuation of the trend that sees the U.S. losing ground

compared to other spacefaring nations, including both established and emerging space powers."

Here are a few facts and analyses from *The Space Report
2012* worth paying attention to, much of the information good
news, but some of it troubling:

- In 2011 there were 84 launches, 14 percent more than the
  previous year; Russia led with 31, China had 19, and the
  United States had 18, marking the first time that Chinese
  launches exceeded those of the United States. The United
  States led in launch vehicle diversity, with eight types of
  orbital rockets launched throughout the year.

- At the end of 2011 there were an estimated 994 active
  satellites in orbit around Earth.

- The U.S. space workforce declined for the fourth year
  in a row, dropping 3 percent from 259,996 in 2009 to
  252,315 in 2010 (the most recent year for which data is
  available); this was the second lowest employment level
  recorded during the previous ten years.

- Average annual space industry salaries were 15 percent
  more than the average salary for the ten science, technology, engineering, and mathematics careers that employ
  the largest number of people in the United States; in 2010
  the average space industry salary was $96,706, more
  than double the average U.S. private-sector salary; the
  states with the highest salaries were Colorado, Maryland,
  Massachusetts, California, and Virginia.

- More than 70 percent of the NASA workforce is between
  40 and 60 years old, with less than 12 percent under age

35, compared to the overall U.S. workforce, where less than 45 percent is between 40 and 60.

- Thirty-four percent of U.S. fourth graders and 30 percent of eighth graders performed at or above the proficient level in science in 2009; 40 percent of fourth graders and 35 percent of eighth graders scored at proficient or higher levels in math in 2011, an improvement over past years.

Changes in Trajectory

Quite a few trends were reported in *The Space Report 2012*, developments likely to affect space activity for years to come. They include changes in the trajectory of human spaceflight; national budget austerity that leads to programmatic uncertainty; increasingly prevalent and diverse partnership models; and the maturing relationship between government and commercial space.

This last trend couldn't be more epitomized than by a milestone reached last year. For the first time, an American private spacecraft was launched and then was docked at the International Space Station. That historic berthing of the automated SpaceX Dragon supply ship was followed by a successful splashdown of the capsule in the Pacific Ocean. It was a powerful message about innovation and private commerce in space and exemplified the clout of NASA-funded U.S. competition to help plug up the country's loss of capacity now missing in action with the retirement of the space shuttle.

The occasion did not go unnoticed by the White House, and they requested my comments. I was glad to oblige, writing:

This week's successful launch and delivery of logistics supplies to the International Space Station by a U.S. commercial space company, reminds us that where the entrepreneurial interests of the private sector are aligned with NASA's mission to explore, America wins. Falcon 9's maiden flight to ISS—and the other commercial space launches that lie ahead—represent the dawn of a new era in space exploration. Nearly 43 years after we first walked on the moon, we have taken another step in demonstrating continued American leadership in space.

Joining me with additional thoughts were leaders in the space community, such as my longtime friend Norm Augustine, retired chairman and CEO of Lockheed Martin. "Successes in commercial space transportation are not only important in their own right," wrote Augustine, "they also free NASA to do that which it does best . . . namely, push the very frontiers of space and knowledge."

Likewise, Bill Nye, CEO of the Planetary Society, drew attention to the event, calling it a huge step and a milestone that enables cheaper, more reliable access to space. "Investments like this, where the private sector and government work together on technical challenges, strengthens our economy by making advanced technology and innovation part of our culture," Nye said. "With the success of commercial partnerships like this, NASA will have the resources to reach farther and deeper into

SpaceX Dragon cargo craft reaches the International Space Station,
October 2012.

the cosmos so that we may all further know and appreciate our place in space."

My Apollo program colleague astronaut Rusty Schweickart helped round out the fanfare, calling the arrival and docking of the Dragon space capsule at the ISS more than historic. "It is, in fact, the beginning of a new era in space exploration, one in which private industry and individual initiative will begin leading the way in the use of near-space activity," said Schweickart, echoing my beliefs. "This is not only exciting and momentous, but is fully in keeping with the American character of risk taking and consequent reward. The long-term results of this 'first' are beyond our ability to see at the beginning of this era, but there is no doubt that it will serve as a huge incentive for young people who now have firm evidence of the value, and opportunity for individual initiative," he added. "Near-Earth space is now firmly a regular part of the human environment along with the air, water, and land. The future is now, once again, opened to imagination, creativity, and dreams!"

I applaud all these comments and see the achievement by commercial rocketeer Elon Musk and his SpaceX team as a first step. Others will follow, cultivating new capabilities that drive down costs and further secure a private-sector toehold in low Earth orbit.

Buzz Aldrin salutes the flag at Tranquillity Base: his proudest moment.

CHAPTER FOUR

DREAMS OF MY MOON

People often ask me to recount my Apollo 11 moonwalking experiences, my reminiscences of being on the moon. When I reflect on that magical, transformative moment in my life, several things jump out at me.

One thing to keep in mind: President Kennedy said send a man to the moon and bring him back safely—*a* man. We could have satisfied that goal by having a person land on the lunar surface, look out the window, maybe deploy a robot, but not open the hatch to the environment. Instead, we chose to have two astronauts moonwalk because of the buddy system.

Thanks to that decision, Neil Armstrong and I stood on the shores of an inhospitable, desolate yet magnificent landscape. Looking at Earth from that perspective, everything I knew and loved lay suspended overhead, residing on a small, fragile, bright blue sphere engulfed by the blackness of space.

What I didn't anticipate until my return to Earth is that America's success in achieving the first landing of humans on the moon was viewed as a success for *all* humankind. Now that's a buddy system! People from every part of the world took pride in collectively declaring, "We did it." Second, in undertaking the Apollo 11 mission, there was a rediscovery of our own precious planet Earth. It's a very special cradle of life that we all reside on.

My stay on the moon is filled with countless other recollections as well.

Once I set foot on the moon, I checked my balance and peed in my space suit's urine collector. I took note that each time I put my foot down there was a spray of dust. And when that dust hit the ground it changed in albedo—in reflectivity and color.

In looking back at that moment in time, putting aside all the pre-mission training, there wasn't a big picture in my mind of the sequence of what we were doing. We did take some pictures walking around the lunar module. We looked for any damage on the *Eagle* and at what the ground looked like underneath our lander. By the way, when I got through walking around the *Eagle,* snapping photos, I gave the camera to Neil. He took most of the pictures. I'm not trying to ease out of any public relations perspective, but we were never briefed on how important the PR pictures would be.

Our stay time on the moon was brief. But the emotion of being first has been long lasting. Still, as we both walked on the moon, I did have the sense of not being as much a member of a team as a follower. If Neil started to do the wrong thing, I wouldn't have

Buzz Aldrin's photo of his own boot print on the moon

known, because I wasn't following a particular order of what we were doing. In some ways, we were thrown out onto the surface and expected to perform a checklist by memory. Set up the flag. Open rock boxes. Put an experiment in place.

So it was very extemporaneous. There was a sense of, "Well, we're here. Let's go do what we're supposed to do. But what is next?" The later Apollo moonwalkers had a little more time to get used to the lunar environment.

One of the strongest sensations I recall is the smell of the moon.

Buzz Aldrin inside the Eagle, *the lunar lander*

Neil and I reentered the *Eagle* lunar lander and repressurized our little home away from home. Lunar dust soiled our suits and equipment, and it had a definite odor, like burnt charcoal or the ashes that are in a fireplace, especially if you sprinkle a little water on them.

Before we left Earth, some alarmists considered the lunar dust as very dangerous, in fact pyrophoric—capable of igniting spontaneously in air. The theory was that the lunar dust had been so void of contact with oxygen, as soon as we repressurized our lunar module cabin it might heat up, smolder, and perhaps burst into flames. At least that was the worry of a few. A late July fireworks display on the moon was not something anyone wanted!

All the official samples collected from the moon's surface were placed in vacuum-packed containers. Neil did grab a contingency specimen. He stuffed it into his thigh space suit pocket, just in case there was a problem that forced us to scurry off the moon in a hurry.

So, following our moonwalks, first I then Neil climbed back on board the lander. That grab specimen was placed on the cylindrical flat top of the ascent engine cover. As the cabin began to fill with air, we both anxiously waited to see if the lunar sample would begin to smoke and smolder. If it did, we'd stop pressurization, open the hatch, and toss it out. But nothing happened. We got back to the business of readying for departure from the moon.

Yes, Apollo 11 was historic, but it was fraught with risks. When we finally set the *Eagle* lander down, with Neil piloting and me calling out descent numbers for him, we had only an estimated 16 seconds of fuel left in the descent stage. On the surface, *if* we had fallen and torn a suit, there wasn't much chance of survival. *If* the one ascent engine didn't ignite or *if* the onboard computer had a glitch, we would never have left the moon. *If* the rendezvous with Mike Collins, circling the moon in the command module, hadn't gone flawlessly, we then would have faced rather nasty consequences. That's just a few of a string of "*ifs*."

Buzz Aldrin deploys exploratory technologies on the moon's surface.

I note that, in recent years, a document has surfaced that was authored by William Safire, then President Nixon's speechwriter, about our Apollo moon mission. It was written, I suppose, in the spirit of considering what if the "if factor" did not work in our favor.

In a July 18, 1969, statement to White House official H. R. Haldeman, Safire titled his internal White House essay "In Event of Moon Disaster" and included this ominous phrasing: "Fate has ordained that the men who went to the moon to explore in peace will stay on the moon to rest in peace."

Calling us brave men, the speech went on to acknowledge that Armstrong and Aldrin know that "there is no hope for their recovery." "In ancient days, men looked at stars and saw their heroes in the constellations," the statement continued. "In modern times, we do much the same, but our heroes are epic men of flesh and blood."

The Safire document added: "In their exploration, they stirred the people of the world to feel as one; in their sacrifice, they bind more tightly the brotherhood of man."

As odd a statement as that sounds today, it didn't surprise me to read it. Speechwriters prepare remarks for all sorts of hypothetical events. Senior officials must always be prepared with remarks for breakthroughs as well as tragedies. Apollo 11 had the potential to fit into either one of those categories. Reading the prepared eulogy, I am proud to say that our mission accomplished the same goals—and brought us back home safely.

Apollo was built on the proficiency and professionalism of thousands of dedicated Americans. It was also built on faith and a national commitment.

By the way, while Neil was the first human to step onto the moon, I'm the first alien from another world to enter a spacecraft that was going to Earth.

A Different Place

The moon is a different place since I traveled there in 1969.

First of all, take your own longing look at the moon in the evening sky. It is obvious that Earth's moon is a celestial body with a story to tell. It has the scars to prove it—a cratered, battered, and beat-up world that is a witness plate to 4.5 billion years of violent processes that showcase the evolution of our solar system.

Thanks to a fleet of robotic probes recently sent to the moon by several countries, there's verification that the moon is a mother lode of useful materials. Furthermore, the moon appears to be chemically active and has a full-fledged water cycle. Simply put, it's a wet moon.

New data on our old, time-weathered moon points to water there in the form of mostly pure ice crystals in some places. For example, sunlight-starved craters at the poles of the moon—called "cold traps"—have a unique environment that can harbor water ice deposits. Gaining access to this resource of water is a step toward using it for life support to sustain human explorers. Similarly, the moon is rife with hydrogen gas, ammonia, and methane, all of which can be converted to rocket propellant.

Fresh findings about the moon from spacecraft have revealed the lunar poles to be lively, exciting places filled with complex volatiles, unique physics, and odd chemistry, all available at supercold temperatures. Recently, the first Lunar Superconductor Applications Workshop was held that brought together expert groups in high-temperature superconductors, low-temperature electronics, cryogenic engineering, and lunar science. The upshot was that even dealing with below 100 kelvin temperatures on the moon, there are several high-temperature superconductors to select from; substances like sapphire and beryllium become thermal superconductors. Digital and analog circuits can operate at very low power and very high speeds, with very low noise and very high fidelity.

It is this kind of harnessing of the exotic that sparks innovation and creativity. What kind of power generation and storage systems can operate for long periods of time that take advantage of the wild swing of lunar temperatures that are available? There's already discussion of lightweight, modular, and expandable superconducting magnets that could provide space radiation shielding on the moon. Those permanently shadowed, incredibly frigid areas at the moon's poles might also be ideally suited for infrared telescope observations.

In short, our celestial neighbor in gravitational lock, the moon, can be tapped to help create a sustainable, economic, industrial, and science-generating expansion into space.

The question is, What should America's role be in replanting footprints on the moon?

Preserving the Apollo Landing Sites

Standing on the talcumlike lunar dust just a few feet from the *Eagle,* the lunar module that transported Neil Armstrong and me to the bleak lunar terrain, I labeled the landscape we stood upon "magnificent desolation."

There were six Apollo lunar landing missions from 1969 to the close of 1972, and just 12 of us were fortunate to kick up dust on the moon. Somebody tagged us the "dusty dozen." The accumulated moonwalking time was limited: From Apollo 11's modest 2.5 hours to Apollo 17's campaign of forays, it added up to a little over 22 hours. Quite literally, exploration of the moon—both robotic and human—has barely scratched the surface in terms of gathering knowledge about that crater-pocked globe.

There's a campaign under way to designate the Tranquillity Base site where Neil and I landed as a national historic landmark. NASA itself has put together recommended guidelines on how to protect and preserve the historic and scientific value of U.S. government lunar artifacts. The field of space heritage preservation is gaining traction.

I'm an advocate for preserving all six Apollo landing sites. By expending the effort to safeguard Apollo 11's Tranquillity Base location, we will learn how best to preserve the other five Apollo landing spots.

I have several ideas on how to proceed. The historic Tranquillity Base landing location could be isolated, encircled by a track system on which movable cameras would be trained on the spot. The lighting conditions would change, given the 14 days of sunlight and 14 days of darkness. Operated by a commercial company with a little creative thinking, it could create an amazing virtual reality experience.

I have looked over the superb forget-me-not images taken from moon orbit, by NASA's Lunar Reconnaissance Orbiter,

The Apollo 11 plaque, now on the moon

that clearly show *Eagle*'s Tranquillity Base landing site. The sharpshooting camera specialist is Mark Robinson at Arizona State University's School of Earth and Space Exploration in Tempe. He is principal investigator for the Lunar Reconnaissance Orbiter Camera.

You can make out the remnants of our first steps as dark regions around the lunar module and in dark tracks that lead to the scientific experiments that Neil and I set up on the surface. There's another trail that leads toward Little West crater, around to the east of our *Eagle* lander. Neil took this jaunt near the end of the two and a half hours we spent moonwalking to steal a look inside the crater. This was the farthest either of us ventured from the landing site. Overall, our tracks in kicking up the lunar dust cover less area than a typical city block.

Robinson observes that hardware launched from Earth and sitting on the moon has been resting there for 40 to 50 years now. Talk about a long-duration-exposure material-sciences

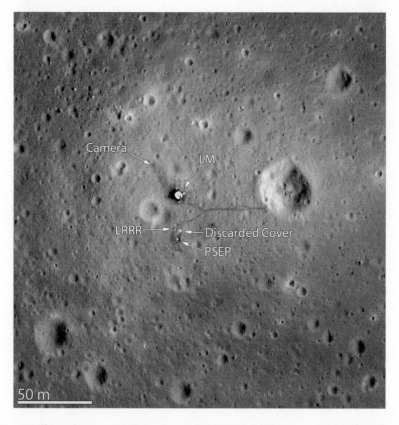

Apollo 11 lunar module landing site (LM), including the Lunar Ranging Retro Reflector (LRRR) and Passive Seismic Experiment Package (PSEP)

experiment, he explains, given radiation, vacuum, temperature cycling, and micrometeorite bombardment.

How have electronics fared? Optics? Paint? Coatings? Metals? Synthetics? In future years, on-the-spot observations and recovery of a modest amount of these materials would be a boon to engineers building lunar hardware, Robinson says. He points out that waste bags tossed out by Apollo crews might make for an interesting biology experiment. Are any microbes still alive among the garbage and human waste left on the moon? If so, can we see evidence of adaptation to the harsh lunar environment?

In May 2012 NASA and the X Prize Foundation of Playa Vista, California, announced that the Google Lunar X Prize, a $30 million competition for the first privately funded team to send a robot to the moon, is also recognizing NASA guidelines to guard lunar historic sites and preserve ongoing and future science on the moon.

I've been wondering if one of those teams might have their robot recover Alan Shepard's golf balls that he hit during his Apollo 14 moon landing mission.

Maybe with all the craters he managed to get a hole in one?

The Bush Push

In January 2004 President George W. Bush put NASA in high gear, heading back to the moon with a space vision that was to have set in motion future exploration of Mars and other destinations. The Bush space policy focused on U.S. astronauts first returning to the moon as early as 2015 and no later than 2020.

Portraying the moon as home to abundant resources, President Bush did underscore the availability of raw materials that might be harvested and processed into rocket fuel or breathable air. "We can use our time on the moon to develop and test new approaches and technologies and systems that will allow us to function in other, more challenging, environments. The moon is a logical step toward further progress and achievement," he remarked in rolling out his space policy.

To fulfill the Bush space agenda required expensive new rockets—the Ares I launcher and the large, unfunded Ares V booster—plus a new lunar module, all elements of the so-called Constellation Program.

The Bush plan forced retirement of the space shuttle in 2010 to pay for the return to the moon, but there were other ramifications as well. Putting the shuttle out to pasture created a large human spaceflight gap in reaching the International Space Station. The price tag for building the station is roughly $100 billion, and without the space shuttle, there's no way to reach it without Russian assistance.

In the end, the stars of the Constellation Program were out of financial alignment. It was an impossible policy to implement given limited NASA money.

Today, pushing the calendar nearly 45 years later, I see the moon in a different light from that of the space race days of 1969. I envision a 21st-century moon, one that can be transformed into an International Lunar Development Authority.

This entity would set the stage for establishment of infrastructure that not only taps the resource-rich moon by commercial, private-sector groups, but also spurs international partnerships between nations.

America can lead the way in creating a lunar consortium of robotic base building that embraces the talents of China, Europe, Russia, India, Japan, and others to establish a firm—this time permanent—foothold on the moon. Moreover, in doing so, the United States can sharpen its own technological know-how that's needed to eventually homestead the red planet.

For several years I have been working shoulder to shoulder with a group of engineers and scientists who are engaged in a vital initiative: an International Lunar Research Base. This base will first be anchored in Hawaii and later evolve to a base on the moon. The project is being carried out under the wing of the Pacific International Space Center for Exploration Systems, or PISCES for short. Its purpose is straightforward. PISCES would

Building up an international lunar encampment

drive the development of surface systems and other hardware for the moon, be it for energy production and storage, recycling, construction, or mining, and spark a host of resource utilization technologies and techniques. Those working on this base have coined a phrase: "Dust to Thrust."

Hawaii and the moon, the coupling of the two brings back memories. In the 1960s, prior to my Apollo 11 flight, NASA made use of the lower slopes of Mauna Kea on the Big Island of Hawaii. It was a training ground for Apollo astronauts, to help us experience what the surface of the moon would be like, and how best to work there. In fact, of all the places on Earth where we trained, the Big Island most felt like the moon.

The proposed International Lunar Research Base can become a unique multinational facility, a test site first on Earth, later to be replicated on the moon. A central goal of this venture is for the United States to acquire the skills for remotely operating robotic systems, knowledge useful to connect habitats, perform habitat-maintenance tasks, set up scientific experiments, and run mobile prospecting gear capable of mining the moon.

Decades ago, there was only one way to put human cognition on the moon. That was the expensive proposition of hurling people and their brainpower there. Today it's no longer the only choice.

Advances in telerobotics can plant human cognition and dexterity on the moon. Telerobotics is an explosive growth industry here on Earth. We plunge to great ocean depths using human-controlled automatons. Robotic equipment extracts resources from perilous mines. Our skies are increasingly dotted with craft that are winging their way under telecontrol. Even high-precision surgery is being done via telerobotics, carried out by a doctor distant from the patient.

Human cognition and dexterity can be extended to lunar territory at the speed of light via telerobotics. Safely tucked inside a high-tech habitat at an Earth-moon Lagrangian point, space expeditionary crews can teleoperate systems that are deployed on the moon.

By demonstrating telerobotic skills at the Hawaii-situated base, processes would be validated in preparation for renewed human activity on the moon. This matchless center will motivate and train the much needed next generation of engineers, scientists, and entrepreneurs primed to take on the challenges ahead in developing the space frontier. I know firsthand, challenging times often precede the most rewarding moments.

First as a terrestrial prototype, a multinational lunar base will help condition us to what's needed on Mars to support future human missions and settlements there.

Cultivating a Unified Effort

When Neil and I stepped upon the surface of the moon at Tranquillity Base, we fulfilled a dream held by humankind for centuries. As inscribed on the plaque affixed to the ladder of our lander: "We Came in Peace for All Mankind." It was, truly, one small step. But more steps are needed. There is no compelling reason to forgo our longer-term goal of permanent human presence on Mars. Consequently, great care must be taken that precious dollar resources needed for the great leap to Mars are not sidetracked to the moon.

The United States has more experience at the moon than any other nation. The country made a huge expenditure in the 1960s and 1970s to gain that leadership. So to just toss that investment away is ridiculous. However, what we now need to do is foster a presence at the Earth-moon L1 and L2 points, libration gateports that permit the United States to robotically assemble, piece by piece, hardware and habitation on the moon. America's space program should help other nations achieve what we have already done.

In chapter 3 I mentioned Lagrangian points—locations in space where gravitational forces and the orbital motion of a body balance each other. French mathematician Louis Lagrange identified these areas in 1772. His gravitational studies of the "Three body problem," suggested that a third, small body would orbit around two orbiting large ones. There are five Lagrangian points in the Earth-moon system, as well as in the sun-Earth system. Because of the combined gravitational force of the two bodies, they can be used by spacecraft as a place to linger, although a spacecraft at the Earth-moon Lagrangian points must use light rocket firings to remain in the same place or control its path around their halo orbits.

The Earth-moon Lagrangian points, E-M 1 and E-M 2, are viable L points: locations where the combined gravity of Earth and the moon permits a spacecraft to be synchronized with the moon in its orbit around Earth. In other words, the spacecraft appears to hover over the far side of the moon. Crew members at this location have continuous line-of-sight visibility to the entire far side of both the moon and Earth.

Gateports between planets will orbit at libration or Lagrangian points.

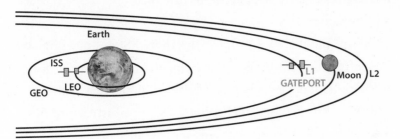

Balanced forces make L1 a key rendezvous point.

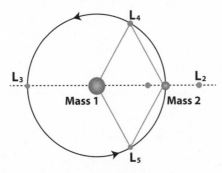

The physics of Lagrangian points

From the Earth-moon L2 point, one scientific setup on the moon is emplacing a far side lunar telescope, equipment that will tune in to an era of the young universe during the first 100 million years of its existence. With no atmospheric distortion, shielded from the buzz and static broadcast from Earth, the extremely "radio quiet" far side of the moon presents a superlative environment for sensitive telescopic observations.

Matching Earth-moon Lagrangian points with astronauts operating telerobotic hardware allows the assembly of infrastructure on the moon, carrying out surface science, scouting out and unearthing important lunar resources. This capability is an innovative advance, redefining the word "exploration"—and it is also a powerful stepping-stone to similar operations at Mars and its moons.

As an initial step, I propose the United States put in place nonsurface lunar infrastructure, including a lunar orbiting global positioning system and libration point relay satellites, as well as space-based fuel depots. These infrastructure projects will enable more efficient and detailed exploration of the moon. For example, a lunar communications system can tackle the challenge of contact with the lunar far side, which is blocked from direct line of sight with Earth. A pair of communication satellites in the halo orbits around the Earth-moon Lagrangian points L1 and L2 would provide radio blackout–free coverage of spacecraft in lunar orbit and for most of the lunar surface.

Available to all countries, the "buy by the byte" lunar communications system would be built to handle an outflow of science data to be returned to Earth, from on-the-prowl teleoperated rovers to robotic sample-return missions that investigate the far side of the moon. First, a lunar communication network will be developed using a frequency common to all users, to be followed by a lunar navigation system.

I've been there. Working on the moon is not easy. You're faced with a lack of reference points and landmarks. The moon is such a small body, the nearness of the lunar horizon makes navigation on the lunar surface tricky. It's very easy to get lost on the surface of the moon, particularly if you are in rough terrain—the very type of landscape that is likely to be most attractive for study.

A lunar navigation system would constitute a constellation of perhaps four or five satellites. They can provide the precise navigation needed to make lunar research much more effective and less risky, both for teleoperated rovers and for human explorers.

This infrastructure is linked to the establishment of a new organization dedicated to cultivating a unified international effort to further examine and develop the moon.

Encouraging Cooperation

Spurred in part by the discovery of lunar water, there has been a major resurgence in moon exploration, carried out not only by the United States. Several nations have their eyes on the moon too, among them China, India, Japan, Russia, and the consortium of countries that form the European Space Agency. Go-it-alone initiatives, though, create the prospect of duplication of effort—and the wasteful use of resources. For spacefaring nations in these turbulent economic times, everyone is dealing with cash-strapped budgets. It is time to build on each other's talents and reduce mission risk by sharing information and capabilities.

How to avoid duplication and make lunar exploration more efficient and more effective?

An International Lunar Development Corporation (ILDC) could be customized to draw upon the legacy and lessons learned from previous efforts, such as the International Geophysical Year, piecing together the International Space Station as well as model organizations like Intelsat and the European Space Agency.

Space collaboration should be the new norm. Despite the Cold War tensions between the United States and the former Soviet Union that characterized the space race of the 1960s, the Russians have become critical partners in the International Space Station—a collective effort of 16 nations. It is time to inspire the international community to jointly explore and develop the moon as a partnership. Forget the space race. That is now a mode that's outmoded.

The ILDC moves beyond dependence on the financial and technical wherewithal of a single nation. What's more, its organizational structure would allow it to easily work with private firms and to make use of private funding. The ILDC should have the flexibility not only to contract with private firms for services and goods, but even to enter into alliances and partnerships with the private sector. Indeed, I visualize the ILDC as an anchor customer for lunar navigation and communications services.

The hallmark of the ILDC will be to encourage cooperation and reduce duplication of effort. Membership will be open to any nation, thus fulfilling the promise of Apollo by allowing all nations a way to take part in the exploration of the moon. For the effective utilization of the moon, goods and services can be bartered, enabling China, India, Japan, and others to land their robotic and human-carrying vehicles. For them, it is about establishing prestige. Space has always been an arena of bartering, and it is a "currency of choice" that has been used in setting up and operating the International Space Station.

So what does the United States barter for? For seats on landers of other nations, U.S. contributions of infrastructure can be offered in exchange for human passenger delivery to the surface of the moon and back. The point here is to avoid investing American taxpayer dollars in transporting U.S. personnel to the moon. We just barter for things and get clever about our ability to negotiate.

America should chart a course of being the national leader of this international activity to develop the moon, but *not* by spending money placing U.S. government people on its surface. There's no need to spend our money on landers and other things that we've done before. Our focus should be limited to robots on the lunar surface that are dutifully employed to do scientific, commercial, and other private-sector work. We need to provide the nonsurface lunar infrastructure and make that available to other governments—China, India, and others—in exchange for an occasional seat on their landers. In short, don't put any more NASA astronauts on the moon!

Our resources must be saved and spent on moving toward establishing human permanence on Mars.

Location, Location, Location

In many ways, I anticipate an outpost on the moon that could mirror what has already taken place here on Earth, in Antarctica.

Americans have been studying Antarctica and its interactions with the rest of our globe since 1956. Visiting researchers delve into glaciology, biology and medicine, geology and geophysics, oceanography, climate studies, astronomy, and astrophysics. Contractors and units of the military provide operational support at year-round stations: Palmer Station, Amundsen-Scott South Pole Station, and McMurdo Station (the main U.S. station

Remote and far away: Amundsen-Scott Station, base camp on Antarctica

in Antarctica). Today, the U.S. Antarctic Program supports a peak population of about 1,600 men and women.

The tab for the U.S. Antarctic Program—an effort that supports scientific research in Antarctica and the waters surrounding, with the goals of fostering cooperative research with other nations, protecting the Antarctic environment, and conserving living resources—is picked up by the National Science Foundation.

On my own trek to Antarctica I could clearly see an analogy between five-plus decades of research in that icy wasteland and what awaits us on the moon. There's a long line of people who hunger to travel to Antarctica and carry out research tasks. Seeking answers to questions leads to new inquiries. The moon is just as complex, as remarkable, and as fruitful in an exploratory sense as Antarctica.

Over the last few years, for instance, a number of different lines of evidence have pooled together to help shore up the case for water on the moon. For example, the Indian Space Research Organization's Chandrayaan-1 spacecraft carried a NASA instrument, the Moon Mineralogy Mapper. It found evidence of water molecules on the lunar surface. The Chandrayaan-1's Moon Impact Probe (MIP) also sensed that it flew through an exospheric "water cloud" during its 2008 plunge onto the lunar landscape. What the MIP found might have been water actually in motion that migrates and concentrates in the ultracold, permanently shadowed lunar craters.

Then there were the NASA Lunar Crater Observation and Sensing Satellite (LCROSS) observations in 2009 that detected water vapor and ice particles kicked up after the LCROSS Centaur upper stage was purposely slam-dunked into the moon.

And here is another appealing link between Antarctica and the moon. Shackleton crater is a large and deep impact feature that lies at the south pole of the moon. This crater was named after Ernest Henry Shackleton, the intrepid Anglo-Irish explorer who took part in the period later labeled as the Heroic Age of Antarctic Exploration, a time span that stretched from the end of the 19th century to the early 1920s.

The shadowed portion of the crater was scanned with the Terrain Camera on board the Japanese SELENE spacecraft in October 2008, helping to gauge the slopes and central peak of Shackleton crater. Those observations were followed by the launch in 2009 of NASA's Lunar Reconnaissance Orbiter. It has played a significant role in eyeing Shackleton with radar and an array of other sensors.

Shackleton crater is more than 12 miles wide and 2 miles deep, about as deep as Earth's oceans. The peaks along the crater's rim are exposed to almost continual sunlight, while its interior is forever in shadow. All this adds up to this captivating feature being an ideal spot for the International Lunar Research Base situated on the edge of the crater. Shackleton hosts both regions of near-permanent darkness and near-permanent sunlight, just the thing for sun-energized power stations. And like real estate here on Earth, it's all about location.

Having water ice within sun-shy Shackleton raises the outlook of harvesting those cold-trapped deposits, an extraterrestrial commodity that would minimize the need to carry water from Earth to the moon. Not only can it be processed for human consumption, it can also be transformed into fuel. Yet another bonus about this crater is that roughly 72 miles away is Malapert

Mountain, a peak that is perpetually visible from Earth and can be topped by a radio relay station.

While there is mounting consensus regarding Shackleton as a future encampment, resolution of the ice issue is likely to require more on-the-spot survey work by robotic craft.

Free Enterprise

As can be hammered out at the Hawaii-based International Lunar Research Base prototype, a crew situated at the Earth-moon L2 position would assemble this permanent facility via telerobotics, piece by piece, module by module. America's return to the moon is one that is robotic, to offer infrastructure and leadership. This pathway eventually spurs private-sector involvement and commercial science that leads to commercial mining. The free enterprise system, if we have a system that's worth its salt, ought to do reasonably well without massive government subsidy. It's American leadership that can create the conditions for commercial development of the moon.

There is a choice to be made. As a country, we can sit around and do nothing. Alternatively, we can take a position of general awareness and accept the role of leadership that we carved out for ourselves in the 1960s and 1970s.

It's very important, in my analysis, to never forget the fact that Apollo affirmed America as a leader in space. Apollo also inspired a new generation to pursue scientific and engineering careers. We should not reengage in a second moon race—we won that contest more than 40 years ago. We should help others in finding their niche in space, while, at the same time, focus on our longer-term goal of permanent human presence on Mars.

Without a doubt, new discoveries about the moon lie ahead.

That, too, is the sense of Paul Spudis, a senior staff scientist at the Lunar and Planetary Institute in Houston, Texas. The moon is close, it is interesting, and it's useful, he observes. As the rocket flies, traversing cislunar space—traveling from Earth to the moon—takes just three days. Additionally, the moon contains a record of planetary history, evolution, and processes unavailable for study on Earth or elsewhere. In terms of its usefulness, projects at the moon can help retire risk for future planetary missions—say sending people to Mars or to the asteroids—by sharpening our space skills and putting to the test exploration hardware for future deep space sojourns.

All this adds up to something Spudis likens to a mantra for moon exploration: "To arrive, survive, and thrive."

Detailed work done by Spudis, gleaned from moon-circling spacecraft instruments that he has helped to develop and operate, reveals that the moon's north pole—at a *minimum*—is home to a large repository of ice. He places it at 600 million metric tons, which, when converted to liquid hydrogen and liquid oxygen, is the equivalent of fuel for a space shuttle launch every day for 2,200 years.

Water is by far the easiest and most useful substance that can be extracted from the moon and utilized to establish a cislunar spacefaring transportation infrastructure. Establishing a permanent foothold on the moon opens the space frontier to many parties for many different purposes, Spudis contends.

By creating a reusable, extensible cislunar spacefaring system, a "transcontinental railroad" in space can be built, connecting two worlds, Earth and the moon, as well as enabling access to points in between.

Spudis and I share a similar perspective. A future lunar outpost can be internationalized, a common-use facility for science, exploration, research, and commercial activity.

Apollo 17's Harrison Schmitt, a geologist and the last man to step onto the lunar surface, has argued for and written extensively about mining helium-3 on the moon to generate economical fusion power on Earth. He advocates a public-private partnership to extract the nonradioactive isotope on the moon.

Veteran space industry entrepreneur Dennis Wingo agrees. He is CEO of Skycorp Incorporated, a small commercial company located at the NASA Ames Research Park at Moffett Field, California.

"Thinking about what can be done with the moon is a lot more practical than complaining about the difficulties," Wingo suggests. He is firmly convinced that the economic possibilities of the moon are great. What remains to be determined is how the moon can be leveraged to solve the 21st-century problems of sustaining and expanding the reach of civilization here on Earth for the nine billion people who will be living here within a single generation.

"It is my firm conviction that the industrialization of the moon is the necessary and logical first goal of the second American space age," Wingo maintains. "The industrial capability of the moon and its near-space environs can now be developed. The industrialization of the moon paves the way for reusable human interplanetary spacecraft, large communications and remote sensing platforms in geosynchronous orbit, and the settlement of Mars."

Building a New Lunar Vision

There are many others who envisage groundbreaking activities on the moon. I can attest to the fact that the moon is a Disneyland of dust. The more time you spend there, the more you get covered from helmet to boots with lunar dust.

But despite its apparent grunge face, the lunar regolith—surface material that's composed in part of rock and mineral fragments—is rich in silicon, aluminum, magnesium, and other useful elements that can be usefully extracted. Two leaders in the use of lunar resources for energy generation on the moon are Alex Ignatiev and Alexandre Freundlich of the Center for Advanced Materials at the University of Houston. Since energy is fundamental to nearly everything that humans would like to do in space, for scientific purposes, commercial development, or human exploration, they are seeking raw materials on the moon that can be utilized to create solar cells on the spot. The moon is an ultrahigh vacuum environment, thus an appropriate setting for the direct fabrication of thin-film solar cells. The lunar vacuum negates the need for vacuum chambers within which to undertake thin-film deposition processing.

The ability to fabricate solar cells on the moon for use on its surface as well as in cislunar space, the researchers believe, can result in an extremely energy-rich environment for the moon.

Ignatiev and Freundlich have looked into the machinery needed to deposit solar cells directly on the surface of the moon. This can be accomplished by the deployment to the moon's surface of a moderately sized cell paver/regolith processor system with the capabilities of fabricating thin-film silicon solar cells. The system could extract needed raw materials from the lunar regolith and prepare the regolith for use as a substrate.

Evaporation of the silicon semiconductor material for the solar cell structure directly on the regolith substrate is done by the paver, with deposition of metallic contacts and interconnects finishing off a complete solar cell array.

This on-moon fabrication process will result in an electric power system that is repairable and replaceable through the simple fabrication of more solar cells, therefore allowing for the expansive use of the moon.

*Circular solar panels and tubular habitations
in a visualization of a lunar outpost*

A power rover could harvest lunar materials for solar cells.

All this is good news for another lunar visionary, David Criswell, now retired director of the University of Houston's Institute for Space Systems Operations. He has long advocated solar power stations built on the moon as a way to provide sustainable and affordable electric power to Earth. The airless moon receives more than 13,000 terawatts of solar power. Harnessing just one percent of that sunlight could satisfy Earth's power needs.

Criswell has promoted a lunar solar power (LSP) system, large banks of solar cells on the moon that collect sunlight. The sunlight is then exported back to receivers on Earth via a microwave beam. That microwave energy is collected on Earth, converted to electricity, and fed into the local energy grid. The LSP can be scaled up on the moon, he contends, to supply the 20 terawatts or more of electricity required by ten billion people.

"The critical frontier for humankind is economic development of the solar energy and material resources of the moon," Criswell concludes.

As I stated earlier, the moon is a far different body today than when Neil and I boot-marked our way across its forbidding face. Scientifically, we know so much more about our celestial next-door neighbor caught in Earth's gravity grip. While there are those who might question the very premise of our undertaking such a journey in the first place, its characteristics were born of the time. It was a Cold War, one-upmanship way to outdistance the former Soviet Union. The moon was the finish line. Apollo was a get-there-in-a-hurry, straightforward space race strategy, and don't waste time developing reusability.

That chapter in the space exploration history books is closed. Today, I call for a unified international effort to explore and utilize the moon, a partnership that involves commercial enterprise and other nations building upon Apollo.

For the United States, other finish lines await.

Asteroids may be rich in resources but may also threaten Earth.

CHAPTER FIVE

VOYAGE TO ARMAGEDDON

There is an important question we all need to face in the immediate future—and that is sustainability. Earth's population is now at over seven billion people. In terms of consumption, the resources obtainable on our globe to provide for that quickly growing number of humans are untenable. At the same time, while we do our utmost to sustain our global security, there are questions about the environmental distress humanity is placing on our planet's ecosphere.

Do we compete for the diminishing resources remaining on Earth—an inward, closed system? Or, alternatively, do we work together in utilizing the limitless resources and opportunities available in outer space—an open and expansive system?

Well, to me, the best option is obvious.

There is history behind events that have menaced life on
Earth. It is also inescapable that there are new threats we cannot
predict. One clear step we can embark on is to enhance the sur-
vivability of our species. We can explore and settle new worlds,
establishing fresh footholds and new beginnings. This is made
possible by evoking bit by bit movement, once again as exemplified by the Mercury, Gemini, and Apollo progression of programs, letting us transit farther into space and accumulate the
know-how to land humans on Mars.

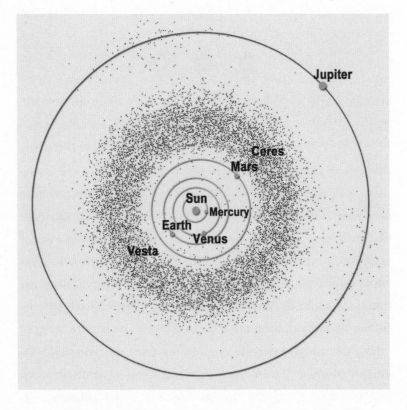

Mapping the belt of asteroids or near-Earth objects (NEOs)

That is precisely what is behind my Unified Space Vision, which preserves U.S. leadership in space exploration and human spaceflight. The USV brings in concert exploration, science, development, commerce, and security elements. That security component I view as one that signifies *both* U.S. defense and a cosmic counterpart: planetary defense of Earth from near-Earth objects, or NEOs.

Planetary defense of home planet Earth means getting to know the enemy—and I am not talking about down-to-earth squabbles between nations. I'm highlighting here a celestial fear factor stemming from asteroids and comets.

NEOs have been nudged by the gravitational attraction of nearby planets into orbits that allow them to enter Earth's solar system neighborhood. We should learn more about these extra-terrestrial wanderers in both scientific and practical terms. I believe my USV essentials of exploration, science, development, commerce, and security fit well with NEOs.

In recasting the U.S. space exploration program in April 2010, President Obama called upon NASA, early in the next decade, to carry out piloted flights to test and validate the systems needed for exploration beyond low Earth orbit. He expected, by 2025, new spacecraft intended for long journeys that would permit America to begin the first ever crewed missions beyond the moon into deep space—starting with sending astronauts to an asteroid for the first time in history. Those deep space assaults are prologue to placing humans in orbit around Mars, returning them safely to Earth, with a human landing on Mars to follow.

In stepping up to Obama's space plan, NASA has begun planning an asteroid mission as the first part of a "capability-driven" approach to explore multiple deep space destinations, acknowledging that the space agency's ultimate destination for human exploration is Mars.

Success requires viable asteroid targets. NASA has identified two accessible space rocks—asteroids 2009 HC and 2000 SG344, NEOs for space travelers to examine in the 2023–2025 time period. But getting there is technology demanding, supported by advanced in-space propulsion, a deep space exploration module that provides adequate habitation for crews, radiation protection, and autonomous operations. A dedicated crewed NEO mission will check out and validate new deep space systems. What is most important, in my point of view, is to ramp up our ability for NEO exploration crew members to perform at demanding destinations while, at the same time, advance our technological skills with each step forward.

So let's look at the exploration, science, development, commerce, and security pieces that are tied to NEOs.

Number one is that near-Earth objects have thumped our world over the ages, and assuredly will in the future. NEOs can shake up but also shape our life-sustaining ecosystem. To assure the survival and guarantee the movement of humanity into space, I feel it is vital we come to terms with NEOs that may have Earth within their crosshairs. Doing so harnesses the technological muscle to not only *encounter* but also *counter* these objects, and it also allows us to use space objects as resource and exploration stepping-stones to Mars, thereby helping to extend the human presence into space.

Earth on the receiving end of a large asteroid

Over the eons, it has been a celestial slugfest. Comets and asteroids have struck Earth since its formation 4.5 billion years ago, bringing seeds of life to Earth early in its history and shattering life by altering the globe's ecosystem, such as hypothesized as the cause for the extinction of dinosaurs.

First of all, I'm not saying you should lose sleep worrying about a giant space rock hitting our world. However, experts that I've listened to advise that, while the chances of a destructive impact here on Earth in the near future are small, they are not zero—and the consequence of a hefty NEO colliding with the planet would be extreme.

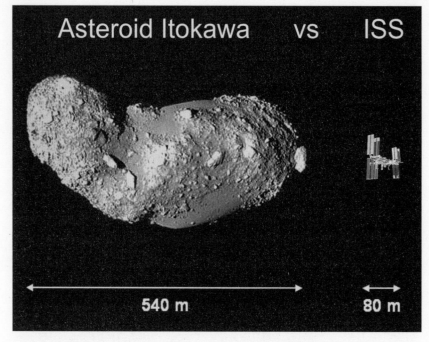

Asteroid Itokawa   vs   ISS

540 m   80 m

*Itokawa, a large asteroid surveyed by Japan's Hayabusa probe,
contrasted with the International Space Station*

For the moment, put aside that mental image of movie star Bruce Willis and his team wrestling with a massive space rock in the hit film *Armageddon*. It turns out that smaller "airbursters" are the more disconcerting sky-slamming flotsam from space. They can cause localized destruction and may infringe upon our air space with surprisingly little warning time.

For example, the 1908 Tunguska event is a saga in which a rocky impactor detonated over remote Siberian real estate, knocking down about 500,000 acres of forest. Supercomputer simulation work led by Sandia National Laboratories principal investigator Mark Boslough suggests that the incoming object was roughly 130 feet in diameter. The object broke up in a cascading way, leading to a rapidly expanding fireball and subsequent blast wave that hit the ground, stirring up a wind strong enough to actually blow over trees. Because smaller asteroids approach Earth statistically more often than larger ones, efforts to detect smaller NEOs would appear to be in order. It is estimated that these smaller objects could impact Earth on average every 2 to 12 years.

More recently, in October 2009, a fireball blast in daylight was observed and recorded over an island region of Indonesia. That atmospheric entry of a small asteroid, perhaps just 33 feet across, rocked their world with a projected energy release of about 50 kilotons, equal to some 110,000 pounds of TNT explosive. Eyewitnesses reported a bright fireball, accompanied by an explosion and a lasting dust cloud.

You can easily visit an impact site of an iron asteroid by traveling to the Barringer meteorite crater, known popularly as Meteor Crater, near Winslow, Arizona. It was formed some 50,000 years ago in flat-lying sedimentary rocks of the southern Colorado Plateau. When that cosmic interloper grooved into Earth tens of thousands of years ago, more than 175 million metric tons of rock were hurled into the sky and redeposited on the crater rim and the surrounding terrain in a matter of a few seconds.

There is an ongoing debate as to the downfall of dinosaurs at the end of the Cretaceous geologic period, 65 million years ago, and the growing consensus is that a mega-asteroid impact caused their mass extinction. That viewpoint stirred up a comment by science-fiction writer Larry Niven: "The dinosaurs became extinct because they didn't have a space program. And

An Arizona crater: remains of an asteroid impact

Siberian forest in ruins: aftermath of a 1908 asteroid

if we become extinct because we don't have a space program, it'll serve us right!" That is insight, and I can't say it much more directly than Niven has.

Getting to Know NEOs

Take note that some 75 percent of Earth is covered by water. What are the consequences if a medium-size asteroid plowed into deep ocean waters?

Research carried out by the late Elisabetta Pierazzo, a senior scientist at the Planetary Science Institute in Tucson, Arizona, served up some bad news. Her work indicated that an asteroid crashing into the deep ocean could have dramatic worldwide environmental effects, including depletion of Earth's protective ozone layer for several years.

There has long been interest in the effects of oceanic impacts of medium-size asteroids, but more focused on the danger of stirring up a regional tsunami. But Pierazzo's approach used computer-modeling scenarios to look at the effects such a strike would have on the atmospheric ozone. The results suggest that midlatitude oceanic impacts of one-kilometer asteroids can produce major global perturbation of upper atmospheric chemistry, including multiyear global ozone lessening.

Pierazzo found that rapidly ejected seawater from an NEO strike includes water vapor and compounds like chloride and bromide that hasten the destruction of the ozone, all of which would influence atmospheric chemistry. Indeed, the removal of a significant amount of ozone in the upper atmosphere for an extended period of time, she found, would have important biological repercussions at Earth's surface—such as an increase in ultraviolet rays that reach terra firma.

So be it by land, air, or sea, getting to know NEOs, I believe, is high on the space program's to-do list. The overall message in terms of planetary defense is that there's need to find them before they find us.

By making use of ground- and space-based technology, humankind does have the ability to anticipate a large-scale impact. Preventing such an occurrence is another matter. Still, to protect life from such a vicious event is an environmental challenge, one that calls upon integrating technology, space policy, and international involvement to launch a global response.

Several fellow space travelers have maintained a long-standing interest in NEOs.

A leader in taking on the NEO challenge is Rusty Schweickart, Apollo 9 astronaut and chairman emeritus of the B612 Foundation. That group announced last year their aim to fund-raise, build, launch, and operate the world's first privately funded deep space telescope mission. Called Sentinel, this project would identify the current and future locations and trajectories of Earth-crossing asteroids. The mission calls for a space telescope—to be built by Ball Aerospace in Boulder, Colorado—to be placed in orbit around the sun, ranging up to 170 million miles from Earth, for a mission of discovery and mapping.

Sentinel appears to be technically sound and on track for a 2017 launch to protect Earth by providing early warning of threatening asteroids. B612 and Ball Aerospace have developed a very viable detection method for finding and tracking near-Earth asteroids. In addition, NASA has forged a Space Act Agreement with the B612 organization to pursue innovative in-space survey skills for detection of new NEO targets.

"For the first time in history, B612's Sentinel mission will create a comprehensive and dynamic map of the inner solar system in which we live—providing vital information about who we are, who are our neighbors, and where we are going," reports Schweickart. "We will know which asteroids will pass close to Earth and when, and which if any of these asteroids actually threaten to collide with Earth. The nice thing about asteroids is that once you've found them and once you have a good solid orbit on them you can predict a hundred years ahead of time whether there is a likelihood of an impact with Earth."

Astronaut Ed Lu, veteran of space shuttle, Soyuz, and space station missions, is the B612 Foundation chairman and CEO. The

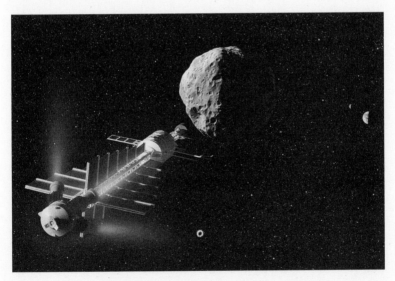

Robotic spacecraft surveys a large asteroid.

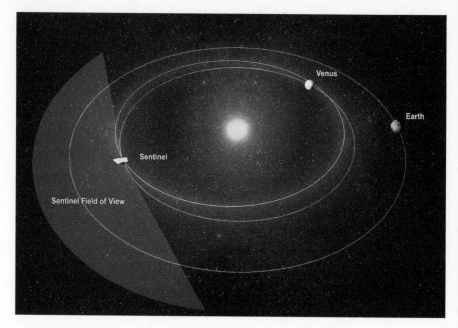

Private Sentinel telescope eyes asteroid population.

extraordinary B612 Sentinel mission extends the emerging commercial spaceflight industry into deep space—a first that will pave the way for many other ventures. "Mapping the presence of thousands of near-Earth objects will create a new scientific database and greatly enhance our stewardship of the planet," Lu believes.

Along with the need to come to terms with the dangers of asteroids, there are several other important outcomes of their study. The United States, Europe, and Japan have successfully hurled spacecraft to asteroids, with more robotic probes to key NEOs on the books.

Russian engineers have been promoting the idea of an automated craft emplacing a location transmitter on asteroid 99942 Apophis, to maintain a more accurate track of this potentially hazardous 690-to-1,080-foot-diameter object. By doing that, we would obtain a very accurate orbit of this NEO, along with an early warning of whether it's on a menacing course with Earth in the years to come.

We already know that the Apophis trajectory places it on an extremely close flyby of Earth in 2029—it is so close, in fact, that it will zip below our geosynchronous satellites. Earth's gravitational tug on Apophis, some worry, may alter its course in such a way as to run into our planet in 2036. But the chance of that happening, experts say, is very, very slim.

My review of Apophis has been enlightening in several ways, specifically in picturing a rotating Earth orbit around the sun where the sun-Earth line is fixed. What an NEO does, if its semi-major axis is *inside* Earth: It does a series of loops around the inside of that circle, then comes back within the vicinity of Earth. Those set of loops are essentially the number of years before it comes back close to Earth.

By Apophis whisking past Earth in 2029, that gravity-assist pass is going to change this NEO so its semi-major axis is *outside* Earth. Until 2036, it will do a series of loops that are outside the circle and traveling in the opposite direction. In other words, Earth's rotating coordinate frame is moving ahead of Apophis.

Simply, Apophis is going to be doing loop-the-loops, getting ahead of Earth, and then it's going to buzz by Earth and do loop-the-loops outside of Earth's orbit. That's what the gravity swingby of planet Earth is doing to this NEO, and I was really amazed when I found that out.

It is a good lesson learned on what an asteroid and its orbit around the sun, or period, do in terms of its availability for a revisit by a robot or a human crew, which is not a constant. It is analogous to understanding some of the inertial cycler orbits that are necessary to sustain a long-term space program.

Getting Our Space Legs

Visitation of NEOs by robotic craft certainly paves the way for human exploration of specific asteroids in the future.

In understanding and coping with the hazard of devastating impacts by NEOs on Earth, we can learn about the physical nature of NEOs. Doing that, in turn, can incrementally enhance our odds of effectively dealing with an NEO, should one of these objects be discovered that could gravely affect us. Furthermore, melding human and robotic abilities at an NEO serves as a test bed to perfect our skills for working at ever greater distances.

In my estimation, human visits to NEOs can go partway toward appreciating the challenges of travel to Mars, without invoking the most severe difficulties. Mars must remain a decisive destination, but NEOs offer a special, practical, and inspiring challenge that gives us the "space legs" to propel deeper toward the red planet.

My research colleague Anthony Genova, at the NASA Ames Research Center, is of like mind. Human exploration of NEOs offers valuable and exciting opportunities as stepping-stones to eventual Mars exploration and colonization. He, too, supports a stepping-stone approach—similar to that seen in the Apollo program—as NEO missions not only reduce the overall risk and complexity of a human space exploration program, but also

NASA scientists simulate an asteroid rendezvous.

decrease the wait time needed for the next "new" mission, allowing the public to lend its crucial support to the program much
earlier than would otherwise be anticipated without intermedi-
ate exploration achievements.

Although asteroids routinely zoom by close to Earth, even
within the moon's orbit, larger and more interesting asteroids
may be tens of millions of miles away. That's a lengthy haul
for people without a resupply of water, food, or air—a mission
longer than has ever been attempted in space and far different from the cargo craft that routinely visit the International
Space Station.

NASA's current goal of having astronauts on an asteroid-
bound mission by 2025 is a core idea promoted by U.S. President

Obama. That call represented a major shift from the space agency's earlier plan, which was aimed at replanting U.S. astronauts on the moon.

Since Obama's 2010 space speech, the interest in transporting astronauts to an asteroid has picked up speed, not only at NASA but also within the aerospace community. The rationale is that such a deep space expedition not only tests out hardware but also builds confidence in humans performing long-duration journeys to other destinations, like the moons of Mars, or onward to the red planet itself. At the same time, a piloted journey to an NEO would provide the savvy to deal with a future space rock found to be on a collision course with Earth.

I term these asteroid explorers "NEOphytes," and they have projected that a human trek to one of those mini-worlds may involve two or three astronauts on a 90-to-120-day spaceflight. The round-trip travel includes a week or two-week stay at the appointed asteroid.

One blueprinted NEO mission, an early human asteroid mission that uses NASA's Orion spacecraft, has been dubbed Plymouth Rock. That plan has been scripted by Orion's builder, Lockheed Martin, and detailed by advanced planner Josh Hopkins.

They portray a six-month mission to an asteroid taking astronauts several million miles from Earth—many times farther away than the moon, but closer than Mars. This requires a very capable spacecraft with propulsion, living space, and life-support supplies, as well as safety features to protect the crew in the event of a problem, since they can't return to Earth quickly.

Frankly, stuffing a crew into the tight quarters of an Orion capsule—even two of them docked together—is not the way to go. Again, I advocate building off of our International Space Station. We need to use our station experience to prototype both a specialized crewed interplanetary habitat and a specialized crewed interplanetary taxi. That's the way to get down to business in projecting ourselves outward into deep space.

What's also urgently required is a much better survey of NEOs, using ground- and space-based assets, to greatly expand the catalog of accessible and meaningful asteroid targets for human exploration. Identification of a sufficient number of accessible and desirable asteroids is critical for future human missions. While the whereabouts of several thousand near-Earth objects

Astronauts will grasp tethers to stay close to asteroids.

131

are known, the number and physical makeup of space rocks that are reachable by piloted flight are highly uncertain. There's a paucity of targets at present to assure maximum mission flexibility. Besides, when it comes to a long-haul, piloted expedition, asteroid size does matter.

Here's my advice: No crew should travel for months on end and pull up to an NEO that's smaller than their own spacecraft! In short, we need to know where to go.

In July 2011 the report *Target NEO: Open Global Community NEO Workshop* was issued, based on a meeting held at

Piloted space exploration vehicles might approach an asteroid.

George Washington University earlier that year. The document pointed out that programs and planned missions to asteroids may be leveraged for mutual benefit in terms of data exchange. It also recommended coordination with the European Space Agency and other space agencies on a planetary defense demonstration mission.

The report points out that a target NEO will need to be discovered several years in advance to provide adequate lead time to deliver robotic precursor missions to scope out the object, plan the human mission, and then send the crew to the chosen objective.

But operating at an asteroid is not a piece of cake. There are great lags in Earth-to-NEO communication times. This kind of deep space mission calls for true autonomy, as crew members are far from Earth, and their space travel must include a great deal of assurance in backup hardware, space propulsion, life-support gear, and radiation shielding. That being the case, a major report finding is that the body of data required to support flying astronauts outward to an NEO is severely limited.

Then there are the psychological and sociological issues linked with an NEO-bound crew cooped up and confined in tight quarters like those offered by an Orion spacecraft. The 2011 report underscores the fact that deep space missions do not afford the abort opportunities and the psychological comfort provided by rapid return to our home planet—a hallmark of my Apollo 11 mission and the six follow-on flights within cislunar space.

My concern here is that far more work is essential to support human expeditions outside of Earth's protective magnetosphere. What still remains as a biological concern is the heightened and long-term physiological effects of space radiation on the human body.

There are arguably many "need-to-knows" about NEOs. That is, just how much data is requisite before a piloted mission departs Earth toward the target space rock? What about the object's spin rate, size and shape, and makeup—solid rock or rubble pile? Also troubling is the ability to station-keep with an asteroid without placing crew and spacecraft in harm's way. In this case, a mobile exploration module deployed from the main spacecraft could carry explorers and robotic tools over to the asteroid. That would seem like a wise and safe approach.

Visits by crews to even the largest asteroids must deal with lack of gravity to safely land. It's more likely to be "docking" to the object in some manner. One idea, proposed by MIT researchers, is tying a lightweight network of tether material entirely around an asteroid. Once in place, astronauts could attach themselves to this set of connections and maneuver or perhaps even walk along the surface. Still, along with the low gravity, asteroids are surely going to be challenging destinations for human and robotic investigation due to the fine, granular topside material spread across the object's surface.

There are ways to make practice runs at NEOs right here on Earth. One technique, which draws upon my early work in underwater simulation of spacewalking, is NASA's Extreme Environment Mission Operations, or NEEMO for short. International crews of aquanauts are trying to understand what a mission to an asteroid would be like. Home base for these

Underwater maneuvers replicate the challenge of asteroid exploration.

evaluations is the National Oceanic and Atmospheric Administration's Aquarius Reef Base undersea research habitat off the coast of Key Largo, Florida, and some 60 feet below the surface of the Atlantic Ocean.

Part of the work is to develop tools and techniques for use on lower gravity environments of NEOs. Working on an asteroid presents special obstacles, say for snagging and bagging geologic samples. Again, great care must be taken as loose material can coast away; an astronaut can be propelled off an NEO's surface just by striking a rock with a hammer.

Cosmic Shooting Gallery

Let's face facts. We live in a cosmic shooting gallery. Ways to defend ourselves from NEOs need careful study. If there is adequate warning time, we have the means to guard Earth from asteroid impacts—a luxury that the dinosaurs were not afforded. But what deflection method to use is still to be determined. There are brute-force concepts, like using a nuclear bomb to blow an NEO to smithereens. Another less harsh option is the "gravity tractor"—a way to alter an NEO's course with a slight nudge over time, using the gravity tug of a spacecraft that has sidled up near the object. Lasers or sunlight-focusing mirrors could also be used to heat up a spot on an asteroid, vaporizing surface material to create a propulsive force that alters the object's path.

There's even been talk of capturing and transporting a small asteroid to near-Earth orbit. A 500-ton asteroid could be fetched and then deposited into a gravitationally stable point in the sun-Earth or moon-Earth system. This NEO moving plan uses a container-like robotic spacecraft powered by a solar electric propulsion system. Once the asteroid is on location, it would be subject to human study, perhaps even a popular tourist stop, as well as exploited as a resource.

Outreach to the asteroids yields a number of benefits. Scientifically, we can find out more about the formation and history of our solar system. From a security standpoint, understanding the structure and composition of asteroids, and learning how to operate spacecraft around NEOs, empowers us to deflect a hazardous intruder from afar. Then there's appraising the feasibility of utilizing asteroid resources for human expansion in space.

Now under way is development of a unique asteroid sample-return mission. This spacecraft is to speed toward 1999 RQ36, a space rock that has the highest Earth-impact probability in the next few centuries of any known asteroid.

NASA's OSIRIS-REx mission is being led by the University of Arizona and is slated to launch in 2016, rendezvous with the asteroid in 2019–2021, then return specimens of the object to Earth in 2023.

OSIRIS-REx is an acronym drawn from the work of the mission: Origins, Spectral Interpretation, Resource Identification, Security, Regolith Explorer. The mission will identify carbonaceous asteroid resources that can be used in human exploration. Another spacecraft duty is to take measurements to quantify the Yarkovsky effect—the daily heating of an object rotating in space can exert a small force on the object.

According to OSIRIS-REx researchers, when the heated surface of 1999 RQ36 points its hot afternoon side in the direction of its motion around the sun, the escaping radiation acts like a small rocket thruster. That propulsive push slows it down and sends it closer to the inner solar system. While that thrust is minuscule, a little push day after day, year after year, for hundreds of years, can alter an asteroid's orbit significantly. More important, the Yarkovsky effect can turn an NEO headed for Earth into an impactor—or a clean miss.

The OSIRIS-REx mission is expected to provide important data, a tool to aid in securing Earth from future asteroid impacts. With time on our side, policymakers can settle on what—if any—steps should be taken to mitigate the odds of 1999 RQ36 banging into Earth.

Pay Dirt!

Extraterrestrial mining in the years to come is one way to spread Earth's economic sphere of influence. Drawing upon the resources of the moon, Mars, asteroids, comets, and other bodies of the solar system can fuel the economic fires of an expanding, outbound civilization.

There are private efforts under way to scope out the job of quarrying space. While the business plans, dollars required, and the technology needed may be jelling, there are thorny questions ahead, issues that organizations are likely to bend their private-sector pick on: property and mineral rights, ownership and possession, international treaties.

The Space Resources Roundtable, often held at the Colorado School of Mines in Golden, Colorado, has increasingly become a hotbed of discussion on these topics. At a roundtable meeting last year, Jim Keravala, Chief Operating Officer of the Shackleton Energy Company, detailed a plan to "fuel the space frontier"—one that would traffic rocket fuel, oxygen, water, and other items into low Earth orbit and on the moon, making this service available to all spacefarers. A mix of industrial astronauts and robotic systems would service customers with a steady stream of propellants and other materials. The business plan calls for liberation of icy resources bound within permanently shadowed craters at the south pole of the moon, processing that material. The company wants to establish a network of refueling service stations in low Earth orbit and on the moon to process and churn out fuel and consumables for commercial and government customers.

But what's ahead is the prospect of legal-beagle debate and court cases concerning mining claims, surface rights, even possession by a squatter. During the School of Mines roundtable, some voiced the point that possession is nine-tenths of ownership. Even the view that it's easier to receive forgiveness than obtain permission circulated among participants.

For a large mining group to get involved in exploiting space resources there must be surety they can make a profit, cautions Dale Boucher, Director of Innovation at the Northern Centre for Advanced Technology, Inc., in Sudbury, Ontario, Canada. He feels that governments should get together and create the regime in which space resource mining can take place.

Most certainly, there are legal matters to be resolved, and before too much is assumed. It's my personal sense that the

United Nations is not the body that should be determining the future legalities of space prospecting and mining. Rather, I see something like an International Orbital Development Authority, an International Lunar Development Authority, and an International Outer Orbit Authority handling these issues.

In the case of setting up the U.S. flag on the moon on Apollo 11, there wasn't a "one small step . . . it's mine" declaration. We set a precedent. We also noted the plaque mounted on the *Eagle* lander that read: *"Here men from the planet Earth first set foot upon the moon July 1969, A.D. We came in peace for all mankind"*—words used to convey that our mission was one of exploration and not conquest.

How sorting out the adjudication of resource-rich celestial objects will play out remains open for dialogue and, quite literally, there is need to dig into these issues deeper.

The outlook for mining asteroids was boosted in 2012 by the intentions of a new private U.S. company, Planetary Resources, Inc., based in Seattle. This team of entrepreneurs announced the venture aimed at mining the solar system, a plan that is billionaire-backed and enthusiastically supported by such people as filmmaker James Cameron, an adviser to the group.

Chris Lewicki, President and Chief Engineer of Planetary Resources, has scripted a multipronged program to access resources from near-Earth asteroids. He makes it clear that developing space resources and creating a market for the volatile mineral and metallic resources of asteroids would be a slice of a larger undertaking. Mining the moon, establishing space-based solar power, and growing a space tourism market are examples of taking the economic sphere of influence on Earth and moving

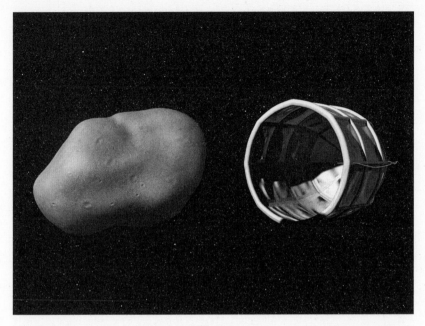

Planetary Resources design for capture of near-Earth asteroid for mining

it beyond the belt of moneymaking geostationary satellites, where it now abruptly stops.

Planetary Resources has outlined a plan to launch a line of low-cost robotic spacecraft. In essence, they have a business plan that calls for the detection, inspection, and interception of asteroids. A first step is to explore for and chart resource-rich asteroids within reach. After intensive study of selected asteroids, the group's intent is to then develop the most efficient capabilities to deliver asteroid resources directly to both space-based and terrestrial customers. What can be extracted from near-Earth asteroids?

Asteroids are floating troves of materials like iron, nickel, and water, as well as of rare platinum group metals—often in significantly higher concentration than found on Earth—such as ruthenium, rhodium, palladium, osmium, iridium, and platinum.

These space rocks vary widely in composition. They can contain water, metals, and carbonaceous materials in various amounts. Some asteroids are loaded with large quantities of water, while other asteroids hold concentrated metals rare on Earth. Water from asteroids is a key resource in space, not only as sustenance for human space travelers but also as rocket propellant.

Certainly not last on the benefit list is furthering American preeminence in space by conducting deep space missions that are practice runs for getting our feet firmly on Mars.

So in summary, prior to conducting either robotic or human missions, securing the target asteroid's orbit and what it's like is critical. For instance, how fast is the asteroid's rotation period; how easy will it be to station-keep alongside the object or "dock" and anchor to the space rock's surface? Surface activities at an asteroid include robotic sample collection and deployment of probes (radar, acoustic, seismometer, et cetera), experiments, and planetary defense devices.

What about the long-duration human interplanetary space mission itself and the unique challenges for the crew, spacecraft systems, and the mission control team back on Earth? Like in the reach for Mars, the drive outward to NEOs needs to utilize the International Space Station to assist in the development of technologies and operational approaches.

Needing emphasis here is that a human mission to an asteroid is a "short-stay" Mars moons mission. It demonstrates, among a list of purposes, linkage to future Mars missions in terms of exercising the transportation system, surveying planetary bodies, furthering deep space operations by crews, and performing tele-operations from a piloted spacecraft to the object being studied.

However the future unfolds, what's needed is a series of steps that convert the NEO natural hazard into natural stepping-stones to support our jump deeper into space. Doing so fills the bill of my Unified Space Vision rules of the road, of exploration, science, development, commerce, and security—and keeps us solidly on the road ahead.

*Buzz Aldrin shows a model of Phobos, one of Mars's moons,
to President Obama.*

CHAPTER SIX

THE MARCH TO MARS

I was an attentive listener when U.S. President Barack Obama declared on April 15, 2010, at the Kennedy Space Center: "By the mid-2030s, I believe we can send humans to orbit Mars and return them safely to Earth."

To fulfill the President's promissory note to the future, I believe that the human reach for the red planet involves a stepping-stone approach, first to Phobos, one of two Martian moons. To be sure, our trips with crews to asteroids prepare us for this rung of the ladder to Mars, as Phobos is like a big asteroid.

Phobos is a way station, a perfect perch that becomes the first sustainable habitat on another world. From that mini-world, crews on Phobos can run robotic vehicles on Mars more directly, in a much shorter communication delay time than commands sent from faraway Earth. Robotic stand-ins for astronauts will ready the habitats and other hardware on the Martian surface, in preparation for the first human crew to arrive on Mars. That's my judgment. My theory right now is that somebody piecing together hardware on Mars through telerobotics on Phobos is the right person to later lead the first landing mission on the red planet.

My approach may well be a contested way to get to Mars, as I'm rubbing up against some unrelenting NASA space planners—but that's not new for me.

Phobos and Deimos are, in a sense, offshore islands of Mars, discovered in 1877 by Asaph Hall at the U.S. Naval Observatory in Washington, D.C. They were tagged with names from Greek mythology: Phobos means "fear," Deimos, "terror." In the future these Martian moons are likely to symbolize just the opposite: courage and security.

Both moons are tidally locked to Mars, as our own moon is relative to Earth: Phobos and Deimos present the same side to Mars all the time.

Phobos is the innermost moon of Mars, only 16.7 miles (26.9 kilometers) in diameter but the larger of the two moons. Diminutive Deimos is a little over 7 miles (11 kilometers) in breadth. Scientifically, both Martian moons are oddballs. There is continual dispute as to where they came from. Just how did they get there? Conjecture about them being captured asteroids or cogenerated with Mars is debatable. These two objects are a cosmic detective story, and we need more clues to sort out their true nature.

Years ago I stirred up a little more than Phobos dust by calling attention to a strange feature spotted on that moon. I termed this oddity a monolith, a very unusual structure. While there are

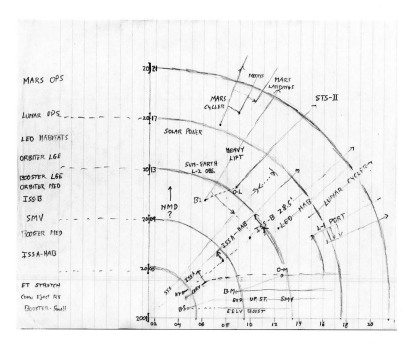

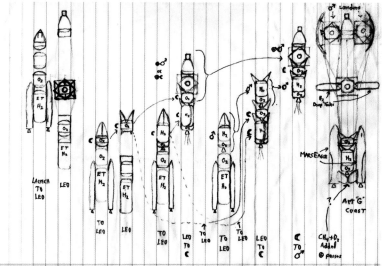

Always inventing: Two of Buzz Aldrin's many sketches. Top, an early time line of missions leading to Mars (2001); bottom, aircraft separation segments preparatory to a Mars landing (2007).

MARS—Western Hemisphere

As with all the names given to extraterrestrial landforms, the features presented in these hemispheric maps carry Latin descriptive terms. The International Astronomical Union (IAU), the official designator, chose that language to encourage scientific discourse and to standardize mapping of the solar system.

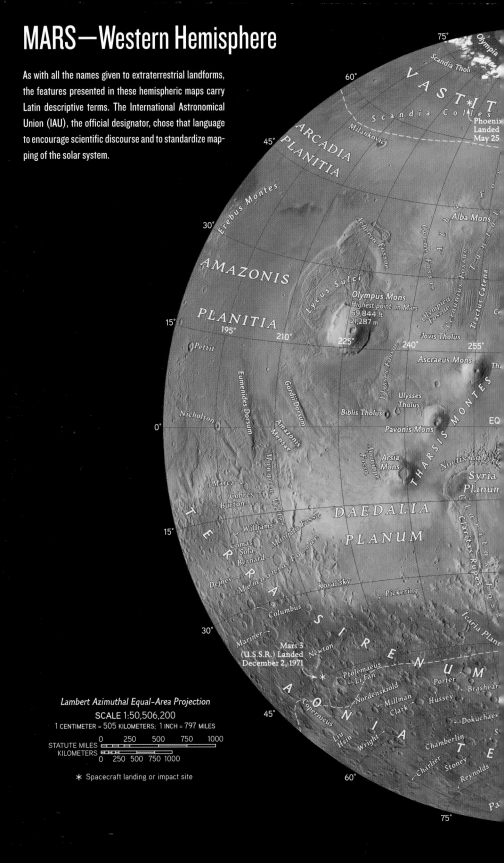

Olympus Mons
Highest point on Mars
69,844 ft
21,287 m

Mars 3
(U.S.S.R.) Landed
December 2, 1971

Phoeni
Landed
May 25

Lambert Azimuthal Equal-Area Projection
SCALE 1:50,506,200
1 CENTIMETER = 505 KILOMETERS; 1 INCH = 797 MILES

STATUTE MILES 0   250   500   750   1000
KILOMETERS 0   250 500 750 1000

✴ Spacecraft landing or impact site

Labels on map:
75° · 60° · 45° · 30° · 0° · 15° · EQ
Olympia
Scandia Tholi
VASTIT*
Scandia Colles
Milankovič
ARCADIA PLANITIA
Erebus Montes
Alba Mons
Acheron Fossae
Cyane Fossae
Lycus Sulci
Olympica Fossae
Ceraunius Fossae
Tractus Catena
AMAZONIS PLANITIA
195° · 210° · 225° · 240° · 255°
Jovis Tholus
Ascraeus Mons
Ulysses Fossae
Pettit
Eumenides Dorsum
Gordii Dorsum
Ulysses Tholus
Biblis Tholus
THARSIS MONTES
Nicholson
Amazonis Mensse
Pavonis Mons
Noctis Labyr
Syria Planun
Marca
Agunopia Fossa
Arsia Mons
Cobres
Burton
Mangala Valles
DAEDALIA PLANUM
Claritas Rupes
Williams
Mangala Fossa
Comas Sola
Bernard
Memnonia Fossae
Dejnev
Kovalsky
Pickering
Icaria Plant
Columbus
Mariner
Newton
SIRENUM
Copernicus
Ptolemaeus
Li Fan
Nordenskiold
Millman
Porter
Brasheal
AONIA
Clark
Hussey
Dokuchaev
Liu Hsin
Wright
Chamberlin
Charlier
Stoney
Reynolds

This color mosaic of the red planet shows how Mars would look to human eyes observing from orbit. Constructed from many thousands of images returned from NASA's Mars Global Surveyor, it details the rocky, desolate terrain of the surface and the distinctive red hue of the planet's regolith. At the poles, frigid ice caps coat the surface, advancing and receding with the Martian seasons.

75°
60°
45°
30°
15°
0°
15°
30°
45°
60°
75°

R E A L I S
Lomonosov
Kunowsky
ACIDALIA
PLANITIA
sonal frost
Perepelkin
Barabashov
is Fossae
T E M P E
num
Sklodowska
CHRYSE CYDONIA
MENSAE
Curie
Nilokeras
Scopulus
Becquerel
T E R R A
Valles
Sharonov
PLANITIA
ARABIA TERRA
SACRA MENSA
Viking 1 (U.S.)
Landed
July 20, 1976
Mars Pathfinder
(U.S.) Landed
July 4, 1997
Trouvelot
Radau
285°
300°
315°
330°
345°
Xanthe Montes
Masursky
Maja Vall
Sagan
LUNAE PLANUM
X A N T H E
Galilaei
us
tes
Orson
Welles
Juventae Dorsa
Mutch
Ophir TERRA
Meridiani
Chasma
Candor Chasma
Planum
Ganges Chasma
Aurorae
Chaos
Opportunity
(U.S.) Landed
Jan. 25, 2004
Planum
alles
Melas
Coprates Chasma
Aurorae
Planum
Dorsa
Chasma
Capri Chasma
Eos
Arima
Beer
V A L L E S
Thaumasia
Vinogradov
IM
M A R I N E R I S
Loire Valles
15°
Melas Fossae
Planum
Nectaris Fossae
Samara Valles
Holden
Mars 6
(U.S.S.R.)
Crashed
March 12, 1974
Bosporos Planum
Bond
Hale
30°
seasonal frost
Nereidum Montes
Hartwig
er
Bosporos Rupes
Vogel
Douglass
Hooke
Arkhangelsky
ARGYRE
Galle Roddenberry
Lohse
Wirtz
Helmholtz
PLANITIA
Charitum Montes
Green
45°
Fontana
Phillips
Maraldi
Darwin
PLANO
Lyell

With the absence of sea
level, elevations are
referenced to a 3,390 km
radius sphere.

See Mars Features on
following pages for
definitions of terminology.

MARS—Eastern Hemisphere

75°
60°
V A S T I T
45°
Micoud
Lyot
Deuteronilus
Mensae
Moreux
Protonilus
Mensae
Colles Nili
Renaudot
30°
Okavango
Valles
Mamers Valles
A R A B I A   T E R R A
Rudaux
Cerulli
Quenisset
Nilosyrtis Mensae
Colloe Fossae
Maggini
Luzin
Nili Fossae
Cassini
Flammarion
Baldet
15°
Gill
Pasteur
Schöner
Antoniadi
15°
30°
45°
60°
75°
Henry
Tikhonravov
SYRTIS
Arago
Capen
MAJOR   Beagle
Janssen
Teisserenc
de Bort
December
0°
Schiaparelli
Schroeter
PLANUM
Meridiani
Pollack
Fournie
Dawes
Oenotria Pla
Planum
Huygens
Oenotria Scop
Mädler
T Y
15°
Flaugergues
Denning
Saheki
Milloch
Marikh Vallis
Bouguer
Cankuzo
Harris
Wislicenus
Schaeberle
Ter
Bakhuysen
Njesten
Lowest point on Mars
• 26 838 ft
8 180 m
Alpheus
Colles
HELLAS PLANI
30°
Le Verrier
Hellespontus Montes
Hellas Chaos
✱ Mars 2
(U.S.S.R.) Crashed
Nov. 27, 1971
Rabe
Amphitrites
Patera
Kaiser
Proctor
Barnard
45°
Maunder
Russell
Malea Planu
Dorsa B
Mitch
Sisyphi
Planum
Holmes
60°
South
PLA
75°

Lambert Azimuthal Equal-Area Projection
SCALE 1:50,506,200
1 CENTIMETER = 505 KILOMETERS; 1 INCH = 797 MILES

0   250   500   750   1000
STATUTE MILES
KILOMETERS
0   250 500 750 1000

✱ Spacecraft landing or impact site

Two notable astronomers, Eugène Antoniadi and Giovanni Schiaparelli, crafted maps of the Martian surface based on their observations in the latter half of the 19th century. They used names from classical mythology, establishing the precedent that the International Astronomical Union came to adopt for Mars and most of the other bodies in our solar system.

75°
60°
45°
30°

REALIS

Panchaia Rupes

Stokes

ydnus Rupes

Mie

Viking 2 (U.S.)
Landed Sept. 3, 1976

OPIA PLANITIA

Phlegra Montes

Hrad Vallis

Galaxias Colles

Hecates Tholus

Adams

Lockyer

Granicus Valles

Hephaestus Rupes

Elysium Chasma

Elysium Mons

Albor Tholus

Phlegra Dorsa

TARTARUS MONTES

Tartarus Colles

Orcus Patera

105°   120°   135°   150°   165°   15°

Amenthes Cavi

NEPENTHES

Eddie

Hyblaeus Dorsa

Amenthes Planum

MENSAE

ELYSIUM PLANITIA

Cerberus

Tombaugh

Hibes Montes

Tholi

TOR

Mars Science Laboratory/
Curiosity (U.S.)
Landed
August 6, 2012

Zephyria Planum

Aeolis Planum

0°

Robert Sharp

Knobel

Gale

Aeolis Mensae

Apollinaris Mons

Lucus Planum

Cerberus Dorsa

Lasswitz

Wien

Herschel

Boeddicker

Gusev

A

Hadley

Al-Qahira Vallis

Spirit (U.S.)
Landed
January 4, 2004

15°

HESPERIA

Graff

Ma'adim Vallis

PLANUM

Müller

Ausonia
Montes

Pal

ERIDANIA

Molesworth

30°

Avarua

PLANITIA

Martz

Greg

Arrhenius

Vallis

THEI

TERRA

Wallace

Kepler

Cruls

Bjerknes

TERRA CIMMERIA

Eridania Scopulus

Extent of seasonal frost

Wells

Planum

Chronium

Campbell

45°

Byrd

Rupes

With the absence of sea
level, elevations are
referenced to a 3,390 km
radius sphere.

Thyles Rupes

Space 2 Probes
(U.S.) Crashed
Dec. 3, 1999

Mars Polar Lander
(U.S.) Crashed
Dec. 3, 1999

60°

See Mars Features on
following pages for
definitions of terminology.

Ultimi
Scopuli

E

75°

MARS'S MOONS

Phobos

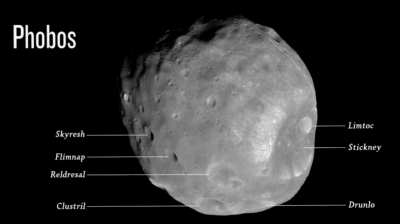

Skyresh

Flimnap

Reldresal

Clustril

Limtoc

Stickney

Drunlo

This irregular moon whips around Mars three times a day, orbiting just 9,377 kilometers (5,827 mi) above the surface. The satellite is only 26.8 kilometers (16.7 mi) across on its longest axis.

Deimos

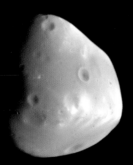

The IAU (International Astronomical Union) has yet to approve names for most features on Deimos.

Mars's smaller moon orbits the planet at 23,436 kilometers (14,562 mi) from the planet. Only 15 kilometers (9.3 mi) in diameter, it completes one orbit of Mars in 1 day, 6 hours, and 17 minutes. Astronomers are unsure of the moons' origins; they may be asteroids captured by Martian gravity, or they might have accreted out of orbiting debris.

MARS FACTS

Average distance from the sun:	227,900,000 km (141,610,500 mi)
Perihelion:	206,620,000 km (128,387,700 mi)
Aphelion:	249,230,000 km (154,864,300 mi)
Minimum distance from Earth:	55,700,000 km (34,610,400 mi)
Maximum distance from Earth:	401,300,000 km (249,356,200 mi)
Revolution period:	687 days
Average orbital speed:	24.1 km/s (15.0 mi/s)
Average temperature:	-65°C (-85°F)
Rotation period:	24.6 hours
Equatorial diameter:	6,792 km (4,220 mi)
Mass (Earth=1):	0.107
Density:	3.93 g/cm^3
Surface gravity (Earth=1):	0.38
Known natural satellites:	2
Largest natural satellites:	Phobos, Deimos

MARS FEATURES
(on maps, preceding pages)

TERM	PLURAL	DEFINITION
Catena	Catenae	Chain of craters
Cavus	Cavi	Hollows, irregular steep-sided depressions usually in arrays or clusters
Chaos	Chaoses	Distinctive area of broken terrain
Chasma	Chasmata	A deep, elongated, steep-sided depression
Collis	Colles	Small hills or knobs
Crater	Craters	A circular depression
Dorsum	Dorsa	Ridge
Fossa	Fossae	Long, narrow depression
Labyrinthus	Labyrinthi	Complex of intersecting valleys or ridges
Lingula	Lingulae	Extension of plateau having rounded lobate or tongue-like boundaries
Mensa	Mensae	A flat-topped prominence with cliff-like edges
Mons	Montes	Mountain
Patera	Paterae	An irregular crater, or a complex one with scalloped edges
Planitia	Planitiae	Low plain
Planum	Plana	Plateau or high plain
Rupes	Rupēs	Scarp
Scopulus	Scopuli	Lobate or irregular scarp
Sulcus	Sulci	Subparallel furrows and ridges
Terra	Terrae	Extensive landmass
Tholus	Tholi	Small domical mountain or hill
Unda	Undae	Dunes
Vallis	Valles	Valley
Vastitas	Vastitates	Extensive plain

Conceptualization of a deep space vehicle, its design in keeping with Buzz Aldrin's Unified Space Vision. Illustrations here and on the following pages by Jonathan M. Mihaly (California Institute of Technology) and Victor Q. Dang (Oregon State University) in collaboration with Buzz Aldrin, Michelle A. Rucker (NASA JSC), and Shelby Thompson (NASA JSC). The vehicle components are based on publicly available concepts for the NASA Deep Space Habitat (DSH), NASA Solar Electric and Cryogenic Propulsion modules (CPS), and the Lockheed Martin Orion vehicle.

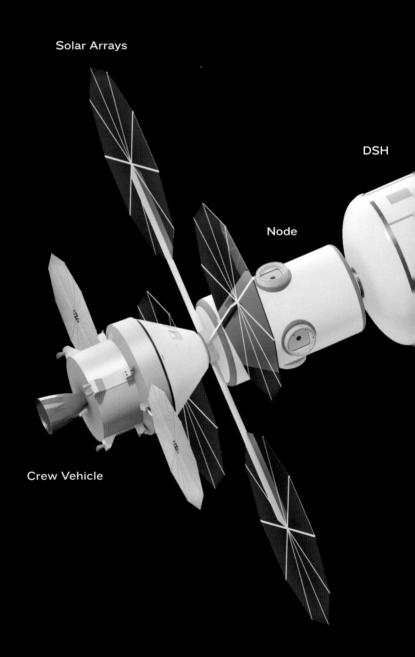

Solar Arrays

DSH

Node

Crew Vehicle

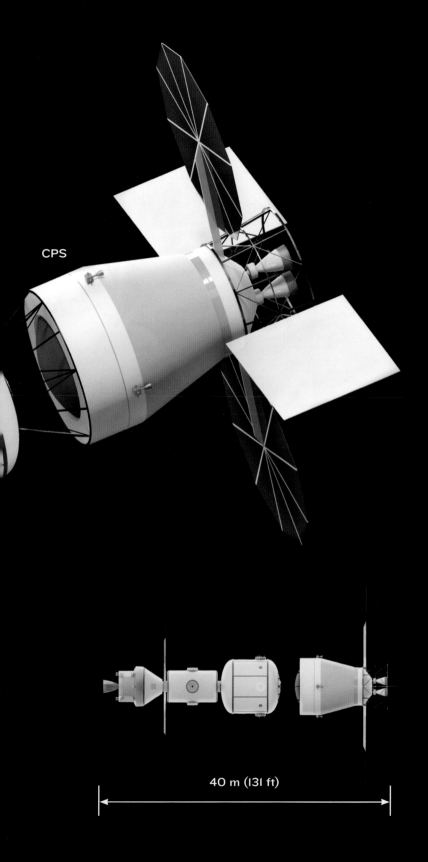

CPS

40 m (131 ft)

To take up permanent habitation on Mars, humans may "terraform" Mars, or turn it more like Earth. This artist's conception includes the following features:

01 Habitation modules, potentially built robotically
02 Pioneer village, designed to extend the stay time of human visitors
03 Global warming processes, to create a water cycle and make the Martian atmosphere conducive to life

04 *Geodesic domes providing climate-controlled living spaces for plants and humans*

05 *and* **06** *Nuclear power plants and wind turbines to generate power for ongoing technologies*

Even with this added infrastructure, human beings will still need supplemental oxygen to live on Mars.

NASA artists have worked with scientists and engineers to envision the equipment and environments necessary for permanent human habitation on Mars.

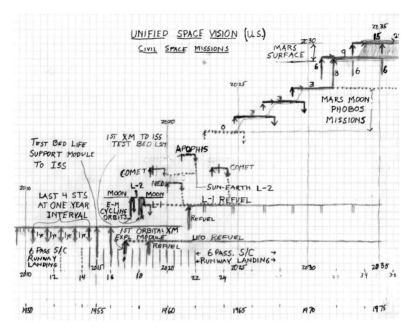

As technology changes, *Buzz Aldrin updates his vision for space exploration. Above, a 2009 time line of his Unified Space Vision. Below, a graphic showing the progress he urges our nation and the world to make in the coming decades.*

BUZZ ALDRIN'S UNIFIED SPACE VISION

In 1903, the Wright Brothers took us into the sky. 66 years later, Apollo 11 took us to the moon. This is how we land on Mars.

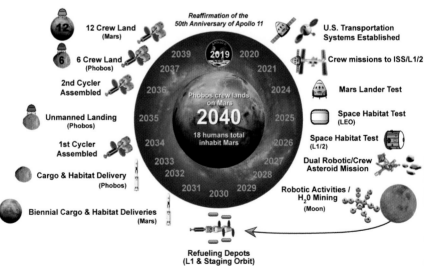

Reaffirmation of the 50th Anniversary of Apollo 11

12 Crew Land (Mars)

6 Crew Land (Phobos)

2nd Cycler Assembled

Unmanned Landing (Phobos)

1st Cycler Assembled

Cargo & Habitat Delivery (Phobos)

Biennial Cargo & Habitat Deliveries (Mars)

U.S. Transportation Systems Established

Crew missions to ISS/L1/2

Mars Lander Test

Space Habitat Test (LEO)

Space Habitat Test (L1/2)

Dual Robotic/Crew Asteroid Mission

Robotic Activities / H$_2$0 Mining (Moon)

Refueling Depots (L1 & Staging Orbit)

2019 · 2020 · 2021 · 2024 · 2025 · 2026 · 2027 · 2028 · 2029 · 2030 · 2031 · 2032 · 2033 · 2034 · 2035 · 2036 · 2037 · 2039

2040

Phobos crew lands on Mars

18 humans total inhabit Mars

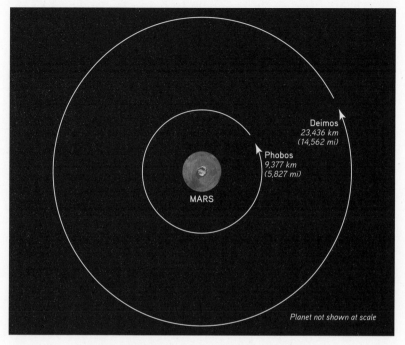

Phobos and Deimos, two moons, orbit Mars.

those who view it as a large, rectangular boulder, visiting Phobos can categorize this curious creation, put there by the universe, or God if you prefer.

The good news here is that Phobos orbits Mars at just 5,827 miles (9,377 kilometers) from the planet's surface. It circles Mars in about eight hours. It is nearer to its parent planet than any other known moon in our solar system. Phobos hurtles around Mars faster than the planet rotates, so future Mars-walkers could see this moon rise and set twice a day. Anyone on Phobos would see how the moon is bathed in reflected light off of the red planet. This "Mars-shine" is akin to earthshine, when sunlight reflects off our planet and illuminates the moon's night side.

147

There is, conversely, long-term bad news for Phobos. Due to its short orbital period around Mars, 50 million years hence it will crash into the red planet, or bust into pieces due to gravitational forces.

Phobos is a heavily cratered, irregular body with no atmosphere. The gravity field is very weak—less than one-thousandth the gravity on Earth—making it easier for spacecraft to land and take off. Escape velocity from this moon is just 25 miles an hour. This moon's most eye-catching feature is Stickney, a six-mile-wide crater. When the object that formed this crater hit Phobos, its impact fashioned streak patterns across the moon's surface. The day and night sides of the moon have been gauged, showing extreme temperature variations; the sunlit side of Phobos is like a pleasant winter day in Chicago, while only a few miles away, on the dark side of the moon, the temperature is more ruthless than a night in Antarctica.

The PH-D Project

Taking all these factors and others into account, I feel that Phobos may well be the ideal location from which to support a "nonhuman, hands-off Mars" program—at least initially. From Phobos a crew can control rovers and other machines to survey Mars and orchestrate the pre-positioning of habitation modules. A Phobos station can draw upon our accumulated know-how in constructing the modular International Space Station. A laboratory bound for Phobos can be certified for duty at the space station prior to send-off to the Martian moon. The regolith of

Phobos can be used to envelop the lab, a way to help protect crews from radiation.

I'm not alone in valuing the Martian moons as fundamental to opening up Mars to human visitation.

Similar in thought is S. Fred Singer, an emeritus professor of environmental science at the University of Virginia. He was the founding director of the National Weather Bureau's Satellite Service Center back in 1962 and has a long pedigree of building and flying space instruments. Moreover, he has advocated his PH-D project for decades, PH-D standing for Phobos-Deimos.

Singer and I were accomplices in early Case for Mars conferences, staged in Boulder, Colorado, starting in 1981 and convened by maverick and passionate members of the "Mars Underground"—motivated largely by wanting to push the throttle forward on reactivating humans-to-Mars planning.

Singer has had an enduring enthrallment with Phobos and Deimos and yet he remains perplexed as to how and why the Mars moons came to be. He favors Deimos as the place to establish a human-tended laboratory. Being higher above Mars, it's easier to get to and is nearly in synchronous orbit, a far better situation from which to observe and operate equipment on the planet below, he suggests.

We agree on the plan for teleoperation of Mars machinery from either Phobos or Deimos. The light-speed distance, even coupled with relay satellites circling Mars, is far shorter than what takes place now between Earth ground control and the NASA Curiosity and Opportunity rovers. Getting closer to Mars permits nearly real-time, fraction-of-a-second teledirection of

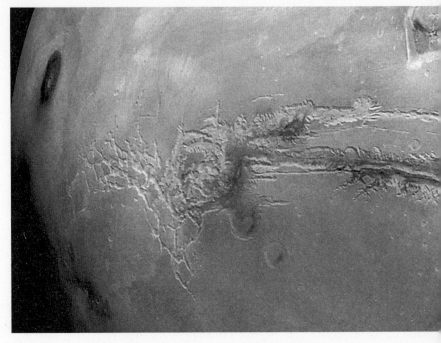

The grandest of canyons on Mars: Valles Marineris

robotic surface equipment on the planet. There's very little delay, within human reaction time.

An added bonus is that the moons of Mars are airless. It's a free vacuum provided by Mother Nature, a real advantage as an environment in which to carry out scientific research on site. For one, specimens collected and rocketed off Mars could be assessed on a Martian moon, thereby curbing forward- and back-contamination worries. That is, you lessen the scene of humans fouling the samples snared on Mars and diminish the risk of nasty Martian biology doing harm to Earth's biosphere when specimens are lugged back to a Mars moon only—in essence, it becomes a bio-barrier between the two planets.

By placing a crew-occupied laboratory/control station on either Phobos or Deimos, an assortment of probes, penetrators, and rovers can be controlled on Mars. Far more of the planet can be reconnoitered, more so than a landed crew could achieve. After all, Mars is vast. It's a huge planet with a lot of real estate, some of it very hazardous in terms of crevasses, caves, steep hills, giant canyons, and high mountains. Better to lose a robot or two than have a person face a deadly predicament.

This is exemplified by new research from the University of California, Los Angeles. In 2012 An Yin, a professor of earth and space sciences, unveiled new data that inform our understanding of plate tectonics on Mars. Using information and images

151

from Mars orbiters, he announced that Mars is at a primitive stage of plate tectonics, pointing to two plates divided by Mars's Valles Marineris, calling them Valles Marineris North and Valles Marineris South. That geologic feature is the longest and deepest system of canyons in our solar system. If the existence of plate tectonics on Mars holds true, it may well bolster the odds that the planet was an extraterrestrial address for life at some point in its past. Therefore, close-up study of this area is warranted, plausibly by low-flying robotic craft that could deploy seismometers, as the site may be rife with landslides, even Mars-quakes.

Setting up a lab/control center on one moon of Mars also allows humans to voyage to the other. This sortie by space taxi would be of great value scientifically, enabling a comparative sampling of both moons. Are they made of the same stuff? Do they have a common origin? As Singer suggests, we simply don't know. Phobos and Deimos are probably the cheapest source of raw materials in the solar system, because the Delta-v penalty is so low. This means that a small, propulsive effort is needed to change from one trajectory to another by making an orbital maneuver. It takes relatively little rocket thrust to transport resources from these mini-worlds due to their small size and, therefore, low gravity.

From a Distance: Tele-exploration

In early August 2012 the one-ton, nuclear-powered Curiosity robot successfully made Mars its home. This car-size rover is NASA's most advanced precursor mechanized system yet, factory

NASA's Curiosity Mars rover takes a self-portrait.

equipped with scientific instruments, cameras, and a robot arm and ready to roll on six wheels for years.

Curiosity's parts are parallel to what a human brings to Mars: body, brains, eyes, arms, and legs. The robot uses antennas for "speaking" and "listening." The one-way communication delay with Earth varies from 4 to 22 minutes, depending on the relative position of our planet and Mars: 12.5 minutes is the average. Curiosity can attain a roaring top speed on flat, hard ground of 1.5 inches a second, equating to some 450 feet an hour.

On one hand, robots are able to cope with the surly climes of Mars while carrying out boring, risky, or dull jobs. On the other hand, humans bring perception, speed and mobility, dexterity, and an inquisitive nature.

Combining the two is opening up a new paradigm in space exploration. "Telepresence" makes use of low-latency communication links that can put human cognition on other worlds. Low-latency yields the appearance of "being there" in a way that is near real-time believable.

The ability to extend human cognition to the moon, Mars, near-Earth objects, and other accessible bodies helps limit the challenges, cost, and risk of placing humans on perilous surfaces or within deep gravity wells.

Let me point out the advances in telerobotics here on Earth. Human cognition and dexterity are already reaching the deepest oceans, pulling out resources from dangerous mines, performing high-precision surgery from a distance—all this as aerial drones, piloted by humans in far-off command centers, fly overhead.

My close friends Robert Ballard and James Cameron can attest to telepresence-enabled undersea exploration, operating

A piloted craft designed for deep-sea exploration faces challenges similar to those of crafts designed for outer space.

vehicles outfitted with high-definition video cameras, sensors, and manipulator arms—run from a mission control. Teleoperation of underwater equipment is also a routine task performed by those maintaining deep-sea oil rigs.

The counterpart in space, albeit showcasing low-quality telepresence, was used decades ago by controllers in the Soviet Union. They wheeled about their automated Lunokhods on the moon. More recently, recall the plucky Spirit and Opportunity Mars rovers run by NASA, precursors to the now-on-Mars Curiosity mega-robot.

Telepresence, low-latency telerobotics, and human spaceflight are leading to redefining what constitutes an "explorer."

A leading champion of exploration telepresence is Dan Lester of the Department of Astronomy at the University of Texas in Austin. Lester tackles the serious concern about how this strategy meshes with our historical concept of "exploration." Telepresence may be effective, and it may be cheap, but if it's not seen as "out there" exploration, it's not going to take hold. Lester's perspective, however, is that putting human cognition in faraway places—if not human flesh, boots on the ground—is a key new capability.

Lester has observed that decades ago when Neil Armstrong and I reached the moon's surface, the only way to put human cognition there was to dig our boots into the ground. That's what we did. But it's no longer the only option.

High-quality telepresence from an Earth-moon Lagrangian point allows a high degree of human cognition and dexterity to be expressed via lunar surface telerobotic surrogates. Lester sees even more significant advantages at Mars, due to the vastly longer two-way latency between Earth and the red planet. Putting humans close enough to an exploration site to ensure cognition—that is, in many respects, what human spaceflight is for.

What is more, telepresence/on-orbit telerobotics is not destination specific. We'll first need to earn our telepresence stripes at the moon and on Mars, using these technologies to explore, scout out mining opportunities, and pre-position habitats without need of on-site, space-suited astronauts. That first Mars base built before human occupancy should not offer sparse living conditions. It should be regal, well thought out, fail-safe; and it should be assembled with care, thanks to distantly operated telerobotics.

Teleoperation at Mars will prepare us. Mars simply tops the list of future destinations to explore. There are plenty of spots ripe for human cognition to encounter, like roaming across hellish Venus and sunbaked Mercury—perhaps even "teleboating" across the liquid ethane and methane lakes of Saturn's moon Titan.

We begin the challenge with our mission to Phobos or Deimos, then Mars.

Red Rocks Mission

Plans for the march to Mars have been percolating within the larger space engineering community. Lockheed Martin has shaped one, based on their Orion spacecraft design. The result is a wished-for undertaking called Project Red Rocks to explore the outermost moon of Mars, Deimos. The aerospace firm sees the proposal as the penultimate stride before boot prints adorn the red planet.

A Project Red Rocks fact sheet from the firm suggests: "Sending astronauts to Deimos will demonstrate key technologies that will be needed for subsequent human Mars landings." The best near-term opportunities to send humans toward Mars, based on Project Red Rocks, would be in 2033 and 2035, thanks to a melding of orbital mechanics, propulsion needs, and a lessening of crew exposure to cosmic radiation. For a 2033 mission, according to company experts, equipment and supplies can be launched in January 2031 and deployed to Mars orbit ahead of time. A Deimos-bound crew would then say goodbye to Earth in 2033, spend 18 months orbiting Mars, and then return to their home planet in November 2035.

Why Deimos? Lockheed Martin space officials see that moon as having a sweet spot, a site near the "arctic circle" on Deimos that offers ten months of continuous sunlight during the Martian summer, enabling the use of simple solar power systems. Astronauts would have direct line of sight to Earth and to rovers on the surface of Mars, simplifying communications, according to the aerospace company.

The view of Mars from Deimos would be stunning. For instance, Olympus Mons alone, the great volcanic mountain on the planet, would be roughly three times wider than the full moon seen from Earth.

Sending astronauts to Deimos will demonstrate key technologies that will be needed for subsequent human Mars landings, such as reliable life-support recycling systems, long-term cryogenic propellant storage, and the biomedical technology to protect astronauts from the effects of microgravity and space radiation, according to Josh Hopkins, principal investigator at Lockheed Martin for advanced human exploration missions. There are things required for the interplanetary trip in space from Earth to Mars and back, he adds, and then there are the challenges specific to actually landing and operating on Mars itself. A journey to Deimos is very similar to the in-space parts of a trip to Mars in terms of distance, duration, and environment.

Project Red Rocks would explore Deimos, the outermost moon of Mars.

The Red Rocks mission would lay the groundwork for landing crews on Mars, contends Hopkins. Planners would have to learn how to guard astronauts from the effects of long-term zero gravity exposure and radiation, build trustworthy water recycling systems, and keep astronauts in high spirits when living in the isolated confines of a small habitat far from Earth.

Library of Alexandria of Mars

Another activist for Phobos and Deimos as stepping-stones for human space exploration is Pascal Lee, co-founder and chairman of the Mars Institute, a planetary scientist at the SETI Institute, and the principal investigator of the Haughton-Mars Project at the NASA Ames Research Center in Mountain View, California.

Lee believes that the Martian moons are emerging as new targets for human exploration, objects that could be visited well before humans reach the surface of Mars itself. Lee and his colleagues have plotted out various science goals to probe the dual moons best done by on-the-scene crews, such as deep drilling and extraction of subsurface samples, 3-D imaging of the interior of each moon via seismic tomography, and searching the regolith of Phobos and Deimos for bits and pieces of asteroids, comets, and maybe the planet Mars itself.

Lee is quick to point out that a human march to Phobos and Deimos can't be supported on science alone. However, if human missions into Mars orbit are part of a logical, stepwise strategy on the way to a human landing on Mars, then the two moons are excellent candidates for the medium term, he says. Furthermore, Phobos, in particular, is ideally positioned to host teleoperated robotic scouts for an in-depth and aseptic reconnaissance of Mars. On that moon, modest infrastructure could be established to process, quarantine, and screen Martian samples brought up from Mars before sending them to eager scientists on Earth.

Lee and his team members suggested several years ago that there's the chance of finding signs of life from Mars ejecta captured by Phobos, a prospect less likely for the outermost moon, Deimos. Consequently, Phobos, he senses, could be the "Library of Alexandria" of Mars. Akin to the ancient Library in Alexandria, Egypt, this Martian moon could likely be a treasure trove,

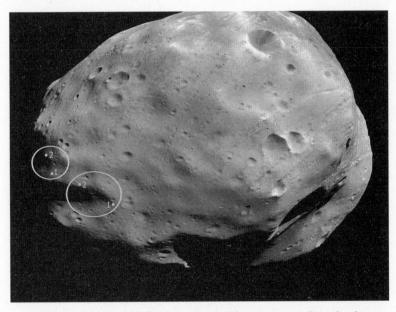

Phobos landing sites were once charted for a Russian robotic lander, but no craft made it there.

rife with knowledge and record keeping that documents all of Martian history.

The idea of finding "little green microbes" on Phobos gained support in 2012 from Jay Melosh, a distinguished professor of earth, atmospheric, and planetary sciences and physics and aerospace engineering at Purdue University. A specimen from the moon Phobos, which is much easier to reach than the red planet itself, he pointed out, would almost surely contain Martian material blasted off Mars from large asteroid impacts. If life on Mars exists or existed within the last ten million years, a mission to Phobos could yield the first evidence of life beyond Earth.

Melosh led a team chosen by NASA's Planetary Protection Office to evaluate if a sample from Phobos could include enough recent material from Mars to carry viable Martian organisms. Combining their expertise in impact cratering and orbital mechanics, Melosh and his associates ran a series of computer simulations.

Their findings support the view that Phobos would have been on the receiving end of Mars material, flung out by large impact events that have happened on the planet over the past ten million years—a relatively recent event in geologic time, in other words. The team plotted more than ten million trajectories and evaluated which would intercept Phobos and where they might land on the moon during its nearly eight-hour orbit around Mars.

When President Obama made his space exploration speech in April 2010, I happened to have with me a small replica of Phobos. I showed the President my Martian moon model, and reiterated my perspective that Phobos is the key to establishing human permanence on another planet in the solar system.

Human adventurers taking root on Phobos is technically achievable. Making use of this moon reduces risk in what must be a step-by-step assault on Mars. There is evolving belief that Phobos enables a steady tempo of exploration and scientific discovery. This moon will not disappoint. As an offshore world of Mars, it allows us to flex our interplanetary muscles, perfect our technological tool kit, and hone astronaut proficiency on the way to our decisive dive onto the beckoning Mars landscape, humankind's future home.

Human habitation and exploration on Mars are within our reach.

CHAPTER SEVEN

HOMESTEADING THE RED PLANET

The red planet has long drawn our curiosity—and now there's a rover prowling about Mars named just that. We first made eye contact with the world that holds its secrets tight thanks to Earth-based telescopes.

Mars is an intellectual magnet provoking thought. Consider the view of astronomer Percival Lowell, writing in his 1908 book, *Mars as the Abode of Life:*

Thus, not only do the observations we have scanned lead us to the conclusion that Mars at this moment is inhabited, but they land us at the further one that these denizens are of an order whose acquaintance was worth the making. Whether we ever shall come to converse with them in any more instant way is a question upon which science at present has no data to decide.

But science about Mars has proceeded ever since, and since 1960, telescopic-driven talk about life on Mars has been augmented by voyages of numbers of automated spacecraft—sent there by multiple nations. Mars has been flown by, orbited, smacked into, radar examined, and rocketed onto, as well as bounced upon, rolled over, shoveled, drilled into, baked, and even laser blasted. Still to come: Mars being stepped on.

Now and in the near future, robotic exploration of Mars is providing a window on a world that can be a true home away from home for future colonists.

The first footfalls on Mars will mark a historic milestone, an enterprise that requires human tenacity matched with technology to anchor ourselves on another world. Exploring Mars is a far different venture from Apollo expeditions to the moon; it necessitates leaving our home planet on lengthy missions with a constrained return capability. Once humans are at distant Mars, there is a very narrow window within which it's feasible to return to Earth—a fundamental distinction between our reaching Earth's moon in the 1960s and stretching outward to Mars in the decades to come.

All this is preface to a major judgment—one that I feel NASA planners are dodging. There is no reason to make a humans-to-Mars program look like an Apollo moon project.

We need to start thinking about building permanence on the red planet, and what it takes to do that. I feel very strongly about this. This is an entirely different mission than just putting people on the surface of that planet, claiming success, having them set up some

Curiosity rover, now on Mars, is a robotic geologist.

experiments and plant a flag, to be followed by quickly bringing the crew back to Earth, as was done in the Apollo program.

What are you going to do with astronauts who first reach the surface of Mars and then turn around and rocket back homeward? What are they going to do, write their memoirs? Would they go again? Having them repeat the voyage, in my view, is dim-witted. Why don't they stay there on Mars?

No question, this is a very big, high-level decision that needs to be made. I can guarantee you, if we have anything like the legislative branch of government in the future that we have today, the first tragedy at Mars with a crew would mean cancellation of the program. And that's all we do about Mars for another century.

I suggest that going to Mars means permanence on the planet—a mission by which we are building up a confidence level to become a two-planet species. At Mars, we've been given

An artist envisions astronauts exploring the Martian landscape.

a wonderful set of moons—two different choices—from which we can pre-position hardware and establish radiation shielding on the Martian surface to begin sustaining increasing numbers of people—not just one select group of individuals. To succeed at Mars, you cannot stop with a one-shot foray to the surface.

It will be a historical moment long remembered when the U.S. President commits the nation to permanent human presence on Mars. Let me hypothesize a political scenario on the 50th anniversary of Apollo 11's landing on the moon, in 2019. The U.S. President, whoever that may be, takes the opportunity to direct the future of human space exploration, pioneered by Americans, by

stating in a speech: *"I believe that this nation should commit itself, within two decades, to establish permanence on the planet Mars."*

That statement will live throughout history, committed to memory on Earth and by the first Mars settlers. In response, around 2020, every selected astronaut should consign to living out his or her life on the surface of Mars.

So why send humans to Mars in the first place?

There is common agreement that humans trump machines in many ways. They offer speed and efficiency to perform tasks. On-the-spot astronauts offer nimbleness and dexterity to go places that are challenging for robots to access. Then there are the innate smarts, ingenuity, and adaptability of a human to evaluate in real time a situation, then improvise to prevail over surprises.

Still, there is a softer side of placing humans on Mars. There are behavior, performance, and human factor unknowns. Living far from Earth in a remote and confined environment will surely induce physiological and psychological stresses. One oddity that is sure to haunt the first humans on Mars—loss of privacy.

You already get a sense of that when you tune in to televised linkups with International Space Station crews. Lots of cameras are positioned everywhere. Of course, the communications time lag between Earth and Mars is a factor. There's a way to start simulating today how best to handle Earth-Mars communications time delays. The International Space Station could simulate and teach people on both ends how to deal with the person-to-person communications delay. What this might boil down to is that every interplanetary traveler has to be a procrastinator!

Vehicles and habitations envisioned by NASA for a Mars post

Other than our limited trips to the moon via Apollo, humans have never embarked upon a mission that's on a par with marching off to Mars; the best analogs so far are Antarctic, undersea, and International Space Station expeditions, but these are distant cousins to the isolation, remoteness, and challenges that will be faced by courageous men and women stationed on Mars, many millions of miles from Earth.

A NASA Mars reference document emphasizes the need for more study of the composition of a Mars crew, based on personal and interpersonal characteristics "that promote smooth-functioning and productive groups, as well as on the skill mix that is needed to sustain complex operations."

Establishing a footing on distant Mars *is* a complex operation. The challenge ahead is monumental and historic. We are on a pathway to homestead the red planet courtesy of robotic explorers that are surveying what now looks like unreal real estate. Nonetheless, there's familiarity with remote Mars. It did not go unnoticed that the first color images transmitted from the Curiosity rover showed layered buttes and other features reminiscent of the southwestern United States.

There is an evolving comfort level with Mars. It is a perspective that beckons us to push forward.

"Retuning" Mars Exploration

I recently took part in a major gathering of Mars aficionados, the Workshop on Concepts and Approaches for Mars Exploration, held June 12–14, 2012, at the Lunar and Planetary Institute in Houston, Texas. It was a heady affair, tackling major issues, including how best to respond to President Obama's challenge of sending humans to orbit Mars in the 2030s.

Some 185 participants came together to share ideas, concepts, and capabilities and address critical challenge areas, focusing on a near-term time frame spanning 2018 through 2024, and a mid- to longer-term time frame spanning 2024 to the mid-2030s.

It was made clear up front that in today's financial times, investment in Mars exploration is tough. The workshop had as a major goal the identification of new concepts and "retuning" ideas for Mars exploration in light of current harsh fiscal realities.

Looking forward into the next decades, all agreed that international collaboration held the greatest potential to enhance future Mars programs—operating in a one-for-all, all-for-one mode. In this light, for instance, any ambitious, complex, and costly Mars sample-return campaign—robotically grabbing and rocketing back to Earth specimens of the red planet—was eyed as dependent upon a long-term and enabling collaboration with other nations.

As always, front and center is the power of Mars to entice us to brood over some key, compelling questions, particularly if life ever was sparked into being there. If so, did it perish or is it still resident on the planet? But also understanding the Martian climate and atmosphere, including the evolution of Mars's surface and interior, can be looped back into grasping the past, present, and future of Earth. The geologic record of early Mars has been preserved, chronicling the period more than 3.5 billion years ago when life is likely to have started on Earth—a time period whose record is mostly missing on our own planet. Can

Mars exploration allow us to turn back the clock and see if life arose elsewhere in our solar system neighborhood?

That's all good science, but we have more urgent reasons to study the environment of Mars: It is mandatory for assuring the safe landing and operation of future robotic and, more important, human missions to the planet. Obviously, unraveling the inner and outer workings of Mars will be a labor-intensive human activity.

There were several challenge areas addressed by workshop attendees, among them what kind of human health-risk reduction is required to support crewed missions around Mars, at Phobos or Deimos, or on the surface of the planet. There's a need to take into account ionizing radiation and the toxicity of soils, among other items.

Analyses of interplanetary trajectories from the vicinity of Earth to the Mars system and return were identified, distinguishing those that offer efficiencies in transportation systems,

An artist's sense of a "space hab" on Mars includes a greenhouse farming unit.

including transit time, cost, et cetera. This includes looking at a variety of Mars orbits and possible rendezvous with or landing on Phobos or Deimos, and scrutinizing trajectories of Earth-moon L2 to the Mars system, and return.

Mars surface system capability is another challenge, whether lighter rover systems able to speed across the planet or equipment that demonstrates in situ resource utilization (ISRU). ISRU demos can shake out equipment to support human surface exploration and settlement—projects for extraction and long-term storage of oxygen and/or hydrogen from available Martian resources in the atmosphere, hydrated minerals on the surface, and digging into Mars to utilize subsurface ice.

Here's an imperative. Incorporating ISRU into human exploration of Mars and its moons necessitates a shift in mind-set—*not* taking everything you need by launching it from Earth. You don't have to haul everything with you because there are available resources at destination's end. There are new ISRU products that can be tapped, such as methane, magnesium, perchlorates, and sulfur. ISRU systems are vital to extract "made on Mars" products from the Martian environment, such as water, oxygen, silicon, and metals for life support, rocket fuels, and even construction materials. Putting in place an effective ISRU system will lessen the need for resupply missions and fully support an off-Earth outpost.

One major realization from the workshop: There is synergy in enabling technologies between robotic and human missions and this increases as future robotic missions become more ambitious. This synergy can manifest itself in a couple of ways, as identified in the meeting:

- Technologies, such as entry, descent, and landing systems, when scaled for application to human missions, enable greater payload mass for robotic missions.
- Leveraging technologies needed for human missions, such as for ISRU and liquid oxygen-methane propulsion systems, can benefit a Mars sample-return mission, due to the potential for reduction in launch and entry mass, hence reducing mission cost.

Of key interest to me, several breakout paper observations produced at the workshop focused on the long-haul vision of preparing for human exploration.

Continual scientific study of Mars is an important prelude to enable targeted, cost-effective human exploration. There's need to extensively characterize the surface and subsurface of Mars. Also, the polar regions of Mars are not only scientifically compelling, they merit study as resource-rich human destinations.

Phobos and Deimos were viewed by workshop participants as "important destinations that may provide much of the value of human surface exploration at reduced cost and risk." It was reported that, as natural space stations and a potential "base camp," these two moons can support teleoperation of payloads on Mars along with habitat buildup, while alleviating some planetary protection issues.

Thanks to robotic surrogates, surprises from Mars will keep coming. NASA's Mars program provided the first close-up photo of the red planet in 1965. Our view of that world has been transformed by camera-snapping orbiters from high above, as well as by the groundbreaking Phoenix lander, a run of rovers—Sojourner,

Spirit, and Opportunity—and now the far more capable Curiosity. They serve as vital precursors for human exploration of Mars—but there is far more work to do.

Flesh and Bone Versus Nuts and Bolts

In striving for settlement of Mars, new technologies must be mastered. Agriculture under extreme conditions, power generation, radiation protection, and advanced life-support systems are called for. Autonomous and highly robust equipment is necessary. To counter the typical 40-minute, round-trip speed-of-light communications time between humans on the surface of Mars and ground controllers on Earth, control must be in the hands of those on the red planet.

Astronauts will use various rovers to expand our knowledge of Mars.

Arguably, one of the better looks at what an early Mars encampment might constitute can be derived from *Human Exploration of Mars Design Reference Architecture 5.0,* issued in 2009 by NASA's Mars Architecture Steering Group. The document was edited by Bret Drake of NASA's Johnson Space Center in Houston, Texas.

NASA's Mars design reference architecture details the systems and operations for a trio of human trips to explore the surface of Mars, carried out over roughly a decade. These first three missions, as the document explains, were designated because the development time and cost to achieve the basic capability to carry out a single human Mars mission "are of a magnitude that a single mission, or even a pair of missions, is difficult to justify."

These first three human Mars missions are also assumed to have been preceded by a series of test and demonstration missions on Earth, in the International Space Station, in Earth orbit, on the moon, and by robotic Mars missions "to achieve a level of confidence in the architecture such that the risk to the human crews is considered acceptable," the report says.

While I differ with sections of this report, it does offer a look at the "need-to-haves" in terms of a starter kit for living on the red planet.

For example, once a crew lands they will need effective and reliable shelter to permit outside excursions. The crew can investigate the Martian surface in a wheeled exploration vehicle, say for weeks at a time, without returning to the habitat. Strolling Mars-walkers will need protection from radiation and dust to safely survey and work on the surface.

From an operational perspective, the first humans to set foot in a Mars landing zone and habitat locale would find themselves at a broad, relatively flat, centrally located area for safety's sake. That means, however, crew and cargo may be far removed from features of scientific significance, beyond a practical walking range for a crew. Pressurized rovers could tote equipment, such as drilling gear to penetrate the Martian surface to moderate depths. The ability to move a drill from location to location would also be desirable. Samples would be returned to the primary habitat that's equipped with a laboratory for extensive analysis.

Not all crew members would trek across the Martian landscape. There would always be some portion of the crew in residence at the habitat.

The NASA report observes that "a strong motivating factor for the exploration of Mars is the search for extraterrestrial life." However, the document goes on to explain that this search could be permanently compromised if explorers carry Earth life and inadvertently contaminate the Martian environment. Additionally, there is need to guard against the remote possibilities that samples transferred from Mars could support living organisms that might reproduce on Earth and damage some aspect of our biosphere. Avoiding both of these eventualities is termed "planetary protection."

One other point—and it's a bit of a catch-22. The fact is that human beings harbor large microbial populations in and on their bodies, and these microbes are constantly reproducing. Even with advances in space suits and habitat construction, it appears impossible that all human-associated activity on Mars *would not* foul the Martian environment. That prompts the rationale that human missions should be sent only to sites where this is tolerable. But that also means avoiding astrobiologically

Exploring Mars: a blend of humans and robots

interesting sites on Mars. Once again, use of sterilized equipment, operated either from a distant, crewed Mars outpost or by astronauts posted at a Mars moon, is likely to be necessary.

Mars researchers Chris McKay and Carol Stoker at NASA's Ames Research Center, along with Robert Haberle and Dale Andersen of the SETI Institute, have long pondered a science strategy for human exploration of Mars. In their view the region containing the Coprates Quadrangle and adjacent areas should be the site of the first human base on Mars. This region is festooned with volcanoes, ancient cratered terrain, and numerous outflow channels; it includes the NASA Viking 1 landing site. That spacecraft settled in on Mars on July 20, 1976, and was the first attempt by the United States at a robotic landing on the red planet.

The main base will occupy the Coprates Quadrangle region, these scientists suggest, and other reachable spots can be maintained as remote field outposts. Orchestrated as an "emplacement

Crew members set up test equipment for polar exploration on Mars.

phase," crews would survey the landing site area to determine the state and distribution of volatiles, especially water. Martian atmospheric gases could be sucked into machinery to supply breathable air for crews, even for cranking out propellants for launching vehicles from the surface of Mars. Resource extraction units would be primed to start stockpiling useful resources. Such stockpiles can provide safety backup to reserves a crew would bring from Earth, supplementing what's available for future arrivals that make their way to Mars.

Deposit, No Return

Long-range thinking has begun on a Mars Homestead Project, one that is identifying the core technologies needed for an economical, growing Mars Base built primarily with local materials.

Bruce Mackenzie, co-founder and executive director of the Mars Foundation of Reading, Massachusetts, along with an active team of like-minded individuals, has plotted out how to build and operate the first permanent settlement on Mars. Some locally derived materials on Mars have been singled out for initial settlement construction, like fiberglass, metals, and masonry, either for unpressurized shelter or covered with Martian regolith to hold the pressurized volume. Polyethylene and other polymers can be made from ethylene extracted from Mars's carbon dioxide–rich atmosphere.

The ultimate goal of the project is to build a growing, permanent settlement beyond Earth, thus allowing civilization to spread beyond the limits of our small planet.

Mackenzie explains there are subtle differences in the technologies required for human settlements on Mars, compared to preliminary human exploration of the red planet. Obviously, duration and reliability of life-support systems is one such difference. So, too, is the need for long-range surface mobility to gain access to a variety of locations on the planet. Lastly, building off experiences gained from the International Space Station, astronauts exploring Mars will need to fabricate hydroponic growth labs where vegetables can be grown. These crops will provide Mars settlers with added nutrition and variety.

An initial goal of a settlement is to build up an infrastructure at one location, Mackenzie reported in 2012 at the 15th annual International Mars Society Convention. "Assuming the settlement is located near the resources it needs, such as an ice deposit, we only need mobility to get to those resources. A variety of spare parts are needed for exploration missions. But a settlement should have manufacturing facilities. Since we can manufacture replacement parts, fewer spare parts are needed."

Mackenzie stressed that the mind-set of Mars homesteaders versus those taking up short-term residence there is completely different. Explorers plan to return to their families and Earth while settlers are there to start a new community and new families. His research conclusion is that there may be noteworthy inefficiencies if we design systems only for human exploration, only to later adapt those systems for settlement. "We should not waste resources developing equipment only used for exploration—other than mobility systems. It would be unfortunate if settlement were delayed forever due to a perceived need to develop technologies which are needed for exploration . . . but not needed for settlement," he suggested.

Those who take the "deposit, no return" voyages to Mars can begin, I believe, to ascertain what can be done in the way of "terraforming" the red planet. That process would alter the face of Mars, intentionally changing its environment to make it a less hostile, highly livable place for humans and to support homesteading the planet. If feasible, and being such a long-term initiative for those on Mars, actions taken must be in concert with informed opinion here on Earth. Specialists, assessing data available, would advise on projected enabling steps if terraforming is to proceed.

The surface area of Mars is equivalent to the land area of Earth. Once a human presence on the red planet is established, a second home for humankind is possible. A growing settlement on Mars is, in essence, an "assurance" policy. Not only is the survival of the human race then assured, but the ability to reach from Mars into the resource-rich bounty of the Martian satellites and the nearby asteroids is also possible. These invaluable resources can be tapped to sustain increasing numbers of Martian settlers, as well as to foster expanded interplanetary commerce and large-scale industrial activities to benefit the home planet—Earth. Of course, some will insist on building outer solar system cyclers as humanity continues bounding into the universe at large.

How Do We Do It?

Mars is key to humanity's future in space. It is the closest planet that has all the resources needed to support life and technological civilization. Its complexity uniquely demands the skills of human explorers, who will pave the way for human settlers.

These words are from Robert Zubrin, a creative astronautical engineer and president of the Mars Society, a group dedicated to further the exploration and settlement of the planet Mars. He is an energetic, effervescent, vocal, and steadfast spokesperson for putting into high gear what he terms the Mars Direct approach—a sustained humans-to-Mars plan that he has scripted.

As author of the pioneering and highly detailed book *The Case for Mars: The Plan to Settle the Red Planet and Why We Must*, Zubrin advocates a minimalist, live-off-the-land approach to space exploration, allowing for maximum results with minimum investment.

Although I differ with aspects of Mars Direct—favoring use of cyclers, pre-placement of Mars habitation modules via teleoperation from Phobos—I applaud Zubrin's spirited nature that is part of a movement that is hastening the day for human settlement of Mars.

Zubrin's blueprint for the red planet uses existing launch technology and makes use of the Martian atmosphere to generate rocket fuel, extracting water from the Martian soil, and eventually using the abundant mineral supplies of Mars for construction purposes. As scripted, the Zubrin plan drastically lowers the amount of material that must be launched from Earth to Mars. That's a key factor to any practical plan for Mars exploration and homesteading.

The general outline of Mars Direct is straightforward, as outlined on the Mars Society's informative website, *www.mars society.org*.

In the first year of implementation, an Earth return vehicle (ERV) is launched to Mars, arriving six months later. Upon landing on the surface, a rover is deployed that contains the nuclear reactors necessary to generate rocket fuel for the return trip. After 13 months, a fully fueled ERV will be sitting on the surface of Mars.

During the next launch window, 26 months after the ERV launches, two more craft are sent up: a second ERV and a habitat module—"hab" for short, which is the astronauts' ship. This time the ERV is sent on a low-power trajectory, designed to arrive at Mars in eight months—so that it can land at the same site as the hab, in the event the first ERV experiences any problems.

Assuming that the first ERV works as planned, the second ERV is landed at a different site, thus opening up another area of Mars for exploration by the next crew.

After a year and a half on the Martian surface, the first crew returns to Earth, leaving behind the hab, the rovers associated with it, and any ongoing experiments conducted there. They land on Earth six months later to a hero's welcome, with the next ERV/hab already on course for the red planet.

With two launches during each launch window—one ERV and one hab—more and more of Mars will be ready for human occupancy. Eventually multiple habs can be sent to the same site and linked together, allowing for the beginning of a permanent human settlement on the planet Mars.

Simulating Mars exploration on Earth

To explore these possibilities, the Mars Society has been running simulated Mars missions in order to test supply requirements, mission hardware, and the ability of crew members to work together under Mars-like settings. Over the years, volunteers have peopled the society's Flashline Mars Arctic Research Station, located on Devon Island in the Canadian Arctic, and a Mars Desert Research Station, set up near the southern Utah town of Hanksville. Other outposts are in position in the Australian outback and Iceland.

The Mars Society's call to attract volunteers to take part in simulated life on Mars scenarios is as direct as Zubrin's plan for settlement of the far-off world: "Hard work, no pay, eternal glory."

Mars Society activists sense, as I do, the untapped reservoir of individuals who value the psychology of becoming a pioneering settler, ready to jump at the opportunity to leave Earth and

SpaceX is developing private Mars operations.

reside on the red planet. History shows us that people *are* willing to risk their lives for great exploits of exploration. Consider the founding of Jamestown in Virginia or the Pilgrims setting foot in Plymouth, Massachusetts—these were daring one-way journeys that led to establishment of permanent settlements.

Why, then, should the call of a New World Mars settlement be any different?

Red Dragon: A Private Affair With Mars

The reach for Mars need not be a governmental event.

One private-sector plan is being led by Space Exploration Technologies (SpaceX), a U.S. commercial space firm birthed in June 2002 by entrepreneur Elon Musk. He gained his fortunes,

Here an artist's rendering depicts Dragon spacecraft on the planet.

in part, from co-founding and then selling PayPal, the online money transfer and payment system.

In May 2012 SpaceX made history when its Dragon space-craft flew atop the company's Falcon 9 booster to become the first commercial vehicle to rendezvous with and then attach to the International Space Station. Dragon is a free-flying, reusable spacecraft under NASA's Commercial Orbital Transportation Services program. This space agency initiative was put in place to spearhead the delivery of crew and cargo to the International Space Station—but turning over these tasks to private compa-nies. The SpaceX Dragon vehicle is made up of a pressurized capsule and unpressurized trunk used for transporting cargo and/or crew members to low Earth orbit.

A restless billionaire, SpaceX's Musk is devoted to taking his Dragon creation to extreme space, breaking away from the con-fines of Earth. His target: Mars.

Under a proposed SpaceX concept, dubbed Red Dragon, the plan is to first send life-looking scientific devices to the red planet using his firm's Falcon Heavy booster. That mission would be followed in later years by sending a human to Mars on a timetable far faster than NASA's. Helping to flesh out techni-cal details of the SpaceX Red Dragon enterprise is a tiger team at NASA's Ames Research Center. They have been sketching out use of the SpaceX craft to search for past or present life on Mars and to sample reservoirs of water ice known to exist in the shal-low subsurface of the red planet.

How this plan evolves over the next few years deserves watching. For the moment, Musk is passionate about the adven-ture and settlement of Mars. Ultimately, it is vital, he says, that humankind be on a path to becoming a multiplanet species. If that human trajectory is *not* pursued, he observes, "we'll simply be hanging out on Earth until some calamity claims us."

A Menagerie of Mars Machines

There are many ways to investigate Mars. A full array of robotic vehicles could be teleoperated by a crew orbiting the planet or stationed on one of its two moons. These same devices might also be deployed and operated by landed crews to boost and expand their exploration presence on the planet.

Remotely piloted Mars gliders and balloons can take to the air. Ground-thumping penetrators, deep-drilling robots, and slithering android snakes could be let loose. Sensor-laden tumbleweed-like vehicles can roll across the planet like dandelions, propelled by Martian breezes to scout about the terrain of Mars using minimum power. Robot hoppers possibly will jump from one spot to another . . . and then another—imbued with a "nose for science," say to use special devices able to sniff the Martian air for traces of biologically produced methane leaking from underground haunts of microbes.

Early on, specially equipped, sterilized automatons will be set to learn more about water on modern Mars. And where there's water, there could be life.

Here is a sampling of inventive and equipment-carrying machines—built to scout out Mars in difficult terrain, hop across the planet from spot to spot, plow into its surface, and even take to the air:

- The *Axel Rover System* is a low-mass robot that can rappel off cliffs and trek agilely over steep landscape. It can look into canyons, gullies, fissures, and craters. Axel can operate both upside down and right side up and is built to scoop up material for later analysis. By using a tether, Axel unreels itself from an anchor point, say from a larger lander or rover, to perform daring descents where humans would find such traverses difficult or too dangerous.

- The *Aerial Regional-scale Environmental Survey of Mars (ARES)* robot aircraft is able to wing its way over Mars. Among its many sky-high duties: Search for possible

Automated vehicles must be designed to investigate challenging features on Mars. A tethered rover might manage steep terrain.

biogenic gases and volcanic gases, measure the Martian atmosphere, and scout out sites for sample-return missions—even help identify spots to land habitats for a future Mars base.

- The *Tracing Habitability, Organics, and Resources (THOR)* project uses projectiles to search out below-surface water ice that may support underground microbial life. THOR aims to use a direct-hit approach to blast out material from beneath the surface of Mars—material that will then be analyzed by an orbiting observer craft.

- *Nuclear-powered "hoppers"* leap from one Martian site to another, examining each locale. An armada of these jumping Mars vehicles swiftly charts large stretches of the Martian surface in just a few years. Hauling science gear from point to point, each hopper sucks up the carbon dioxide–rich Martian atmosphere for use as propellant. On cue, stored heat from a radioisotope power source hits the propellant and shoots the hopper in an arcing path toward a new landing area.

On the Books: MAVEN and InSight

At NASA and elsewhere, sending future robotic missions to Mars remains a cash-strapped activity. However, two NASA spacecraft have been funded to depart Earth for Mars in 2013 and in 2016, respectively. One is an orbiter, the other a lander, and both will add to our reservoir of knowledge about the Mars of the past and how the planet fits into our future.

The Mars Atmosphere and Volatile EvolutioN (MAVEN) mission is designed to survey the planet's upper atmosphere, ionosphere, and interactions with the sun and solar wind. The goal of MAVEN is to unravel the role that loss of atmospheric gas to space played in changing the Martian climate through time. Where did the atmosphere—and the water—go? Basically, this orbiter is to probe how Mars turned hostile.

In circling the red planet, MAVEN's sensor suite will determine the loss of volatile compounds—such as carbon dioxide, nitrogen dioxide, and water—from the Mars atmosphere. That inquiry will give scientists a way to look back into the history of Mars's atmosphere and climate, gauge its liquid water status, and disclose just how the planet appears to have become increasingly inhospitable for life.

Selected in August 2012, NASA's InSight mission to Mars is scheduled for departure from Earth in 2016. InSight's snappy name stands for Interior Exploration using Seismic Investigations, Geodesy and Heat Transport—and that says it all. InSight will get to the "core" of the nature of the interior and structure of Mars.

Is the core of Mars solid, or liquid like Earth's? Data collected will help scientists understand better how terrestrial planets form and evolve. Carrying sophisticated geophysical gear, InSight will delve beneath the surface of Mars, detecting the fingerprints of the processes of terrestrial planet formation, as well as measuring the planet's "pulse" (seismology), "temperature" (heat flow probe), and "reflexes" (precision tracking).

Once on Mars, this craft will take the heartbeat and vital signs of the red planet for an entire Martian year, two Earth years.

The ARES robot aircraft can test the chemistry of the Martian atmosphere.

The THOR project plans to use projectiles to search Mars's surface.

Proposed nuclear-powered hopper jumps between sites on Mars.

Drilling underneath the red Martian topsoil, InSight makes use of a stake called the Tractor Mole, within which an internal hammer rises and falls, moving the stake down in the soil and dragging a tether along behind it. The German-built mole will descend up to 16 feet below the surface, where its temperature sensors will judge how much heat is coming from Mars's interior, and that reveals the planet's thermal history.

The InSight lander is outfitted with a seismometer to take precise measurements of quakes and other internal activity on Mars. Radio signals sent between InSight and Earth will allow researchers to precisely gauge the wobble of Mars, a technique to judge the distribution of the red planet's internal structures and better grasp how the planet is built.

Building the Escalator to Mars

My approach for homesteading Mars is via the Purdue/Aldrin Cycler. First of all, keep in mind two terms when considering this transportation system: There are cycler trajectories and cycler vehicles.

I have long admired and worked with James Longuski, professor of aeronautics and astronautics at Purdue University. Along with his colleagues, we have forged ways to launch a substantially large vehicle that would provide radiation shielding and spacious quarters in order to guarantee the safety and comfort of outbound-to-Mars and inbound-to-Earth astronaut crews.

Cycler trajectories are the paths that cycler vehicles travel on. In many ways, they can be thought of as the highways on which space vehicles travel. Cycler trajectories are routes used over and over again on paths around the sun. These trajectories are identified by using the laws of celestial mechanics—essentially Newton's laws.

Interestingly, to create 21st-century sustainable space transportation architecture I'm counting on laws of motion compiled by Sir Isaac Newton in his work *Philosophiae Naturalis Principia Mathematica,* first published in 1687. Newton's laws of motion have led to a trio of physical laws that form the basis for celestial mechanics. They describe the relationship between the forces acting on a body and its motion due to those forces. My cycler design depends on these principles.

The Aldrin Cycler is a cycler trajectory that travels around the sun, making close flybys of Earth and Mars, a trajectory that takes $2\frac{1}{7}$ years to complete and then repeats every succeeding $2\frac{1}{7}$ years. If a vehicle is launched into the Aldrin Cycler trajectory, it would continuously shuttle between the two planets forever, without requiring a significant amount of propellant to keep on track.

The cycler vehicle does not stop when it flies by Earth. The astronauts have to board a small but speedy space taxi that catches up with the cycler. The cycler is like a bus that repeats its route over and over, but never stops. As a future space traveler you'll have to run fast to catch up and get on the bus!

But once the astronauts are on the cycler vehicle, they can relax and enjoy the ride to Mars. When they arrive at Mars they must board a small vehicle that makes a fiery entry into the atmosphere of Mars. If the astronauts do not get off at Mars, then they will travel back to Earth, getting off $2\frac{1}{7}$ years after they first left Earth.

*NASA's Mars Atmosphere and Volatile EvolutioN (MAVEN) will explore
the upper atmosphere of the red planet.*

By using the Aldrin Cycler trajectory it takes less than 6
months to get to Mars. However, any astronaut not disembark-
ing at Mars would spend 20 more months getting back to Earth.
My Purdue University associates have identified Aldrin Cycler
trajectories that make a short trip—6 months—to Earth from
Mars, and a long trip—20 months—to go from Earth to Mars.

Therefore, a complete Earth-to-Mars human transportation
system would include two cycler vehicles, one using the "outbound" or "up escalator" trajectory to get to Mars and the other
using the "inbound" or "down escalator" trajectory.

Once these cycler vehicles are built and placed in orbit about
the sun, they will continue to freely travel back and forth.

A future Mars lander—InSight—will investigate the internal geology of the red planet.

Propellant will be required, however, now and then, to keep the Aldrin Cycler going—but the cost of refueling is not prohibitive.

What are the biggest challenges?

The Aldrin Cycler requires very high rendezvous velocities at both Earth and Mars—typically 6 kilometers a second (over 13,400 miles an hour) at Earth and as high as 10 kilometers a second (22,370 miles an hour), or more, at Mars. Those speeds make it very difficult for the space taxis to catch up. Think of it this way: If a bus were going 5 miles an hour, riders could easily jump on, but not if the speed was 50 miles an hour!

Can anything be done about the high rendezvous speeds? Yes.

My Aldrin Cycler idea has inspired the search for other Earth-Mars cycler concepts. For example, there are "low-thrust" cyclers that use electrical propulsion to reduce the approach speeds. There are also "four-vehicle" cyclers that take $4\frac{2}{7}$ years to complete their trajectories. Then there are powered three-synodic-period cyclers that require three cycler vehicles. There is even a one-vehicle cycler.

All of these new cyclers are spin-offs of my original Aldrin Cycler thinking, and all have much lower flyby velocities. Each has its advantages and disadvantages. As always, economics is a factor as more vehicles mean more cost. Overall, powered and low-thrust cyclers will demand advancements in propulsion technologies—but this type of progress is well within reach.

So, how shall we go to Mars?

The best, most effective way is still under intensive review. But I'm happy to report that the Purdue/Aldrin Cycler and its offspring will continue to be an important mission design concept in the future development of an Earth-Mars transportation system for human space travel.

Neil Armstrong and Buzz Aldrin erect U.S. flag on the moon.

CHAPTER EIGHT

THE CLARION CALL

Humanity is destined to explore, settle, and expand outward into the universe.

But doing so urgently requires a rekindling of America's space program. A Unified Space Vision can ignite a new wave of support and participation in the United States and elsewhere. This is a spot-on space trek of inspiration, one that can impel youth to engage in science, technology, engineering, and math. Younger readers have probably heard their parents or grandparents say: "The world is yours." I want to take it one step further and say: "The worlds are yours."

When I was a young person, I wasn't the only one in the neighborhood who looked upward and dreamed about going to the moon or stepping onto other planets. I was a reader of science fiction. At that moment in time, no one had traveled into space. Everyone, including me, had to bank on imagination to conjure up ways to make those dreams come true.

To the younger reader, Will you be one of the first people to walk on Mars? You could even be among the first human settlers to colonize that planet. There are out-of-this-world things to accomplish—all fostered by the ability to reach for places that no one has ever reached for in the past.

Earth isn't the only world for us anymore.

The space voyages beyond our Earth over the next 25 years will also motivate the next wave of technology entrepreneurs. This search for new horizons will enhance America's global leadership and encourage international cooperation among spacefaring nations.

Homesteading our solar system is a reach outward to what lies beyond—and beyond to the stars.

I was deeply saddened by the passing of my good friend, and space exploration companion, Neil Armstrong, in 2012. As Neil, Mike Collins, and I trained together for our momentous Apollo 11 expedition, we knew of the technical challenges we faced as well as of the magnitude and weighty implications of that historic journey. We will now always be connected as the crew of the Apollo 11 mission to the moon in 1969. Yet for the many millions who witnessed that remarkable achievement for humankind, we were not alone. An estimated 600 million people back on Earth, at that time the largest television audience in history, watched Neil and me walk on the moon.

Whenever I gaze at the moon, I feel like I'm in a time machine. I am back to that precious pinpoint of time, nearly 45 years ago,

The Apollo 11 crew—Buzz Aldrin, Neil Armstrong, and Michael Collins— celebrate the mission's 40th anniversary in July 2009.

when Neil and I stood on the forbidding, yet beautiful, Sea of Tranquillity. We both looked upward at our shining, blue planet Earth, poised in the darkness of space. I now know that even though we were farther away from Earth than two humans had ever been, we were simply the spearhead of a community of participants. Virtually the entire world took that unforgettable journey with us.

With Neil's death, I was joined by many millions of others from around the world in a global mourning for the passing of a true American hero and the best pilot I ever knew. It had never come to my mind that our Apollo 11 mission commander might be the first of us to pass.

203

My friend Neil took the small step but giant leap that changed the world, and he will forever be remembered as the person who represented a seminal moment in human history.

I had truly hoped that on July 20, 2019, Neil, Mike, and I would be commemorating the 50th anniversary of our moon landing. Regrettably, this is not to be. Neil will most certainly be there with us in spirit. Surely, if we had all been together, we would have collectively supported the continued expansion of humanity into space. Our small mission that was Apollo 11 helps make that possible. But like our fellow citizens and people from around world, we all will miss this foremost aviation and space pioneer.

Neil did not see Apollo 11 as an ending. Rather, he saw our touchdown at Tranquillity Base as a first small step for humankind into the cosmos. He was truly a gifted engineer, consummate astronaut, and leader. Yes, he was soft-spoken and reserved, advocating quietly for space exploration from behind the scenes. He didn't seek fame or honor for the work that he knew so many others had done to make our moon landing achievable.

May Neil's vision for our human destiny in space be his legacy. As he once observed, there are still "places to go beyond belief."

We both expressed that sentiment during periodic visits to the White House, where we discussed U.S. space policy with a succession of Presidents. Conversation in some cases turned to where the next step into the future should lie: Return to the moon or on to Mars? For me, Mars. Neil disagreed. He thought that the moon had more to teach us before we pressed onward to other challenges. Still, while we disagreed at times on that next destination and how best to get there, we were both resolute and shared a common belief: America must lead in space.

Neil's passing was also a time to reflect on those who gave their lives in pursuit of making real the dream of space exploration: the astronauts of Apollo 1 and of the space shuttle orbiters *Challenger* and *Columbia*. We can honor them all—and the U.S. President who set in motion the moon-landing challenge before the country—by renewing our commitment and resolve to space exploration, and pursue it with the same fortitude and durable commitment to excellence that was personified by Neil Armstrong.

Continue the Journey

My call is to the next generation of space explorers and their leaders. It is now time to continue that journey, outward past the moon.

The three of us on Apollo 11 traversed the blackness and vacuum of space to win a peaceful race with a very capable competitor, the former Soviet Union. Apollo 11 was, at its core, about leadership. A young American President challenged himself—and all of us—to think daringly and not withdraw from a shared vision of what we could do in space. The path that John F. Kennedy motivated us to choose was, indeed, not easy. In truth, it was very hard and audacious in scope. But it served the betterment of America, and ultimately contributed to ending the Cold War.

It was an honor and privilege to be a part of the Apollo 11 mission and share in that pinnacle moment within the U.S. space program. There are those who look back at that time and ask,

Inspiring the next generation with a visit to Germanna
Community College in Culpeper, Virginia, November 2014

What did it mean that America was first on the moon? The right question to ask, however, is, What can America do now to build upon that accomplishment decades ago?

Apollo 11 was rooted in exploration, about taking risks for great rewards in science and engineering, about setting an ambitious goal before the world—and then finding the political will and national means to achieve it. Even today the voyages of Apollo seem incredibly bold. Looking back at that time, we are continually stirred by the enormity of the endeavor, one that was energized by the teaming efforts of people from all walks of life, from industries big and small, who worked in tandem to attain a long-term goal of magnificent achievement.

The crew of Apollo 11 was backed by hundreds of thousands of American workers, the greatest can-do team ever assembled on the face of Earth. That team was composed of scientists and engineers, metallurgists and meteorologists, policymakers and flight directors, navigators and suit testers, and those on the shop floor, such as the seamstresses who stitched the 21 layers of fabric for each custom-tailored space suit. They devoted their lives and professional energies, minds and hearts, to our mission and to the other Apollo expeditions. Those Americans embraced commitment and quality to surmount the unknowns with us.

All of these lessons are worth learning anew today. Yes, we live in difficult times. We face these challenges together.

I believe that valiant strides forward in space not only reflect our country's greatness, but summon us to make discoveries that, in turn, improve our lives on Earth. I also sense that national leadership and a coming together of the American people are the ingredients that make overcoming obstacles possible. Apollo 11 is a symbol of what a great nation—and a great people—can do if we work hard, work together, motivated by strong leaders with vision and resolve.

What Apollo 11 means to us today is realizing the dream of exploration by way of determination—and it is a message we need to carry forward into our future.

My vision for our space future is founded on the Apollo tradition. But this time, there is no moon race. Rather, I see the moon as a true stepping-stone to more stimulating and habitable destinations. The moon should act as a new global commons for all nations as we venture outward to Mars for America's future! It is not outside our reach.

That future is already being cultivated as U.S. space entrepreneurs are opening up the space ways to tourism for hundreds of ordinary individuals. This future I envision builds upon the International Space Station, which should become an orbiting research center for all nations—including India, China, and other countries that aim to explore space.

It is a future in which we travel to Earth's orbit aboard new reusable spacecraft, successors to the now retired space shuttle fleet. These are multipurpose international and commercial ships capable of runway landings and supporting an assortment of space duties.

The space station is the ideal test bed for long-duration life support. We must use our station know-how to prototype specialized safe-haven, interplanetary, and taxi modules, hardware that can be combined with Orion-type crew vehicles for missions that cycle back and forth between Earth and the moon. These cyclers are eventually stationed at lunar vicinity, those special places beyond low Earth orbit that I've written about earlier, the Earth-moon L1 and Earth-moon L2 slots. Once in position, they relay communications and double as refueling depots. Making use of cyclers, we also fly by comets and intercept asteroids, particularly the menacing to Earth space rock Apophis.

This is merely a snapshot of what is practicable.

Who knows what the future may bring, what's right around the technological corner or what new revelation in physics is yet to be found? Propulsion via gravity waves, space elevators on the moon, satellite power beaming from point to point in space, our first contact with extraterrestrial intelligence? All of this is part of our 21st-century pursuit of new knowledge and the ongoing process of discovery.

By implementing a step-by-step vision—just as we did with the single-seater Mercury capsule and two-person Gemini spacecraft that made Apollo possible—we will plunge deeper and deeper outward. On the agenda of solid stepping-stones in space exploration: multination and commercial use of our neighboring moon, several human landings on Phobos, the inner moon of Mars. Those exploits are prelude to our historical and milestone-making commitment to homestead the red planet itself.

If collectively we have the vision, determination, support, and political will—and Apollo clearly showed us that these elements can be tied together—then these gallant missions of exploration are within our grasp.

I have a message in a time bottle for the candidate who wins the 2016 election for the U.S. presidency.

There's opportunity to make a bold statement on the occasion of the July 2019 50th anniversary of the first humans to land on the moon: *"I believe this nation should commit itself, within two decades, to commencing American permanence on the planet Mars."*

Making that declaration will be predicated on answering a set of questions:

America, do you still dream great dreams?

Do you still believe in yourself?

Are you ready for a great national challenge?

I call upon our next generation of space explorers—and our political leaders—to give an affirmative answer: Yes!

209

Looking beyond the moon to the asteroids and Mars

BEYOND THE HERE AND NOW: THE RISK AND THE REWARD OF EXPLORING SPACE

I remain steadfast in my view that Earth isn't the only world for humanity anymore.

The first human footfalls on Mars will flag a historic milestone. Success in that effort will ensure a continuously expanding human presence elsewhere in the cosmos, one that I believe will depend on American global leadership in uniting all evolving, capable nations in space exploration.

That said, such leadership is not attainable—much less sustainable—if NASA's budget support continues for years to come at the meager rate of only one half-cent of every U.S. taxpayer dollar. I join those who recently petitioned the White House to change this appallingly insufficient commitment to America's space program—a program that has motivated so many; yielded so much technology; and spotlighted time and time again the country's leadership, ingenuity, and big thinking.

The reach outward to attain a U.S.-led human permanence on Mars requires the necessary funds for the next three to four decades, beginning with an immediate increase of NASA's funding to one percent of the federal budget, followed by a graduated increase to carry out a presidentially and congressionally endorsed objective of landing the first humans on the red planet.

But all these steps must be predicated on one major decision. And the countdown is on for a U.S. President willing to seize the most ideal instant in time to announce that decision. That time is the 50th anniversary of humanity's first touchdown on the moon in July 1969. In 2019, five decades after that triumph of technology and human spirit, he—or she—can utter these words: *"I believe this nation should commit itself, within two decades, to achieving an America-led, permanent presence on the planet Mars."*

Sound familiar? These Kennedyesque words from long ago need to be evoked again, to relight the can-do spirit of space exploration—an unwavering ambition that leads to the first humans residing on Mars.

In touring around the globe to help endorse this book I'm often asked: "One-way trips to Mars? That's a big pill to swallow!" Although the initial Mars crew that establishes "living room" for others might return to Earth, I strongly back the international acceptance that travelers plant themselves on the red planet for good. There are several economic reasons for not bringing them back.

For one, to return people from Mars requires a fueled ascent stage on Mars's surface and an accessible transportation link back to Earth—a major and costly add-on to any plan to populate Mars. The initial and ongoing cost per person for short stays on that planet is far higher than the cost of supporting a crew committed to a permanent presence on that world.

Let's face facts. Billions and billions of dollars are going to be invested by multiple nations in the selection and transport of the first humans to set foot on Mars. And you want to bring them back? Why, and what would they do here on Mother Earth? Given advanced communications technologies on Mars and those here on Earth, everyone can visually and viscerally watch a new branch of humanity take hold on Mars.

We must see in the stars a destination and destiny and make new commitments for long-term human permanence off Earth. We must begin to foster an interplanetary citizenry.

As outlined in this book, *Mission to Mars: My Vision for Space Exploration,* what's called for is a step-by-step vision to plunge deeper and deeper outward.

In many ways, the International Space Station (ISS) is already, and should—along with other space stations—continue to be, a hub for promoting the human flow to the moon and Mars. Crucial life-support equipment can be advanced there. Onboard tests of humanlike robots, telepresence experiments, 3-D printing technology—these have already found a niche on the ISS. Additionally, the station's purpose can be augmented to prototype the interplanetary hardware, exploration modules, and landers that will safely, reliably, and routinely transport crews to the far-off shores of Mars. Space station test bed investigations should assist in the fabrication of future fuel depots positioned at or near Earth-moon Lagrangian points. Ideally, ISS will serve as a crucible of creativity to set in motion my idea of cycling spaceships, an indispensable ingredient to make it feasible and efficient to reach and stay put on Mars.

Meanwhile, expanded and meaningful American cooperation in space with other nations—particularly including China, as well as India, Russia, and South Korea—is paramount. It adds up to human spaceflight for peaceful purposes. That pursuit could help all parties see a universal future and gain perspective beyond present conflicts, and it sets the stage for a common presence, first on the moon and then on Mars.

What about a U.S. comeback at the moon? I remain adamant that America's participation there must be limited to robots that are human controlled from a distance. Those automatons would perform duties like leveling the uneven lunar terrain—at both nearside and farside locales—and would enable the coupling and tending of lunar exploration modules landed there by various countries. I see it as a "partnership for progress"—one that involves commercial enterprise and other nations building upon the Apollo inheritance. This capability on the moon can be honed early on, practicing the techniques and technologies on the Big Island of Hawaii. In fact, pathfinder work is already under way there. More important, the linked efforts at sharpening our space skills here on Earth, on the space station, then in crew rotations on the moon will serve as a proving ground for how best to apply those abilities in homesteading Mars.

I'm heartened to say that since *Mission to Mars: My Vision for Space Exploration* was first published, in 2013, individuals at Purdue University, the Massachusetts Institute of Technology, the California Institute of Technology, the University of Texas, and Stanford University have stepped up, eager to flesh out elements of both my Unified Space Vision Institute and Strategic Space Enterprise. Similarly, the prestigious International Academy for Astronautics is prepared to assess my plans. The outcomes from careful scrutiny of my space strategies will be sent to NASA and perhaps implemented as the next White House shapes its space policy agenda. Urgently needed, I feel, is a focus upon competitive domestic heavy-lift space launch and payload users.

Another activist role I'm undertaking is amending the 1994 bill that designated early astronauts as honorable Space Emissaries. It ought to include the 24 Americans on successive missions that reached the moon, and their title should be changed to Lunar Ambassador. At its heart, this bill remembers all the celebrations spotlighting the individuals and milestone-making missions they flew in the 1960s and 1970s. We must remember the historic cadence of space exploration, the Mercury, Gemini, Apollo, Skylab, and Apollo-Soyuz programs. There's a need to remind everyone of the rapid rate of progress that was achieved that led to placing humans for short stays on the moon.

This same information should be shared around the world, but in a spirit of cooperation. I'm determined to see the United States regain a strong leadership role in low Earth orbit and on the moon, not by competing with other nations but by cooperating and assisting them—all in preparation for the activities needed to set up a permanent outpost on Mars. All spacefaring nations will do well to recall the stepping stones along the way to the flagship event of July 20, 1969, observing as a world united the 50th anniversary of humankind's first moon landing, and then throttling forward by initiating the next concerted giant leap to Mars.

I would be remiss by not noting that 2014 was marked by great global success in space—and by catastrophe, too.

On the achievement side, Europe's Rosetta mission to a comet successfully deployed Philae, the first robotic craft in history to land on a celestial wanderer. India succeeded in reaching Mars with its own scientific orbiter. China circumnavigated the moon and parachuted back to Earth a robotic spacecraft as a precursor test to a lunar sample return program—practicing key elements, I contend, for eventual human sojourns to the moon.

NASA's Mars Atmosphere and Volatile EvolutioN (MAVEN) spacecraft reached the red planet and began its unprecedented task of studying the Martian atmosphere. Meanwhile, down on the planet, the Curiosity rover has been in nonstop mode in its quest to assess whether Mars ever was, or is now, a cozy habitat for extraterrestrial life-forms. Additionally, the robot is helping guide plans for missions of exploration by humans intent on homesteading that planet.

Then there's the highly productive search for other planets beyond our solar system. Ground- and space-based observations have discovered hundreds of planets orbiting other stars. Advanced techniques are in the offing to further survey newly found extrasolar worlds for evidence of intelligent life elsewhere in the universe.

On the human spaceflight side, NASA's Orion spacecraft— built to enable crews to reach Mars and other deep space destinations—flew in unpiloted mode for the first time. The Orion capsule was launched atop a Delta IV Heavy booster from Cape Canaveral Air Force Station in Florida. It made a two-orbit, four-hour flight that tested many of its systems before a high-speed reentry back to Earth, safely parachuting into the Pacific Ocean, where it was met by awaiting recovery teams.

But 2014 also saw failure and tragedy.

En route to resupply the International Space Station, the commercial Orbital Sciences Antares rocket suffered a catastrophic failure just a few seconds after takeoff. The misfortune destroyed its Cygnus cargo-carrying spacecraft as wreckage fell onto the launch area, the Mid-Atlantic Regional Spaceport at NASA's Wallops Flight Facility in Virginia.

There was a disastrous midair breakup of the commercial suborbital Virgin Galactic SpaceShipTwo. The loss of one of its two test pilots underscored the fact that opening up the space frontier is a hazardous enterprise. My good friend Richard Branson and the Virgin Galactic team will assuredly press ahead. And so too are other companies that are keen on shaping 21st-century space businesses and passenger-carrying operations, such as XCOR Aerospace; Blue Origin, which was set up by Amazon .com founder Jeff Bezos; Sierra Nevada's Dream Chaser; and Elon Musk's SpaceX, a visionary enterprise with the ultimate goal of enabling people to live on other planets.

Regrettably, sacrifice has been and will continue to be a truism for spacefaring pioneers. Over the decades, we have lost numbers of individuals—several of them close personal friends of mine—all intent on pushing the boundaries of exploration and seeking new horizons.

The future is sure to present many more missteps as humanity pioneers the space frontier. Risk and reward is the weighing scale of exploring and taming space.

It is time that America examines our goals in space—and recommits to the ones that make the most sense. In summary, history does not make itself. It is made by actions, but also by inaction.

Americans will either set the course in space again and make history or let others do so. My personal appeal is to the current U.S. President and those who follow—as well as to all global leaders. Once again, we are at an inflection point in human history.

We must look up and gather strength from vision and commitment to worthy goals beyond ourselves . . . and beyond the here and now.

—*Buzz Aldrin*
Satellite Beach, Florida

APPENDIX

CHANGING VISIONS FOR SPACE EXPLORATION

✸ ✸ ✸

A TIME LINE OF PRESIDENTIAL POLICIES and actions with highlights from key speeches on space exploration from the middle of the 20th century on.

Dwight D. Eisenhower (in office 1953–1961)

President Dwight D. Eisenhower was President when the Soviet Union launched the world's first artificial satellite, Sputnik I, in October 1957. This seminal event shocked the United States, started the Cold War space race between the two superpowers, and helped lead to the creation of NASA in 1958.

However, Eisenhower didn't get too swept up in the short-term goals of the space race. He valued the measured development of unmanned, scientific missions that could have big commercial or military payoffs down the road.

For example, even before Sputnik, Eisenhower had authorized a ballistic missile and scientific satellite program to be developed as part of the International Geophysical Year project of 1957–58. The United States' first successful satellite, Explorer I, blasted off on January 31, 1958. By 1960 the nation had launched and retrieved film from a spy satellite called Discoverer 14.

John F. Kennedy (1961–63)

President John F. Kennedy effectively charted NASA's course for the 1960s with his famous speech before Congress on May 25, 1961, repeating its bold promise in Texas the next year.

The Soviets had launched Sputnik I in 1957, and cosmonaut Yuri Gagarin had become the first person in space on April 12, 1961, just six weeks before the speech. On top of those space race defeats, the U.S. plan to topple the Soviet-backed regime of Cuban leader Fidel Castro—the so-called Bay of Pigs invasion—had failed miserably in April 1961.

Kennedy and his advisers figured they needed a way to beat the Soviets, to reestablish American prestige and demonstrate the country's international leadership. So they came up with an ambitious plan to land an astronaut on the moon by the end of the 1960s, which Kennedy laid out in his speech.

The Apollo program roared to life as a result, and NASA embarked on a crash mission to put a man on the moon. The agency succeeded, of course, in 1969. By the end of Apollo in 1972, the United States had spent about $25 billion on the program—well over $100 billion in today's dollars.

Special Message to the Congress on Urgent National Needs

Speech before a Joint Session of Congress, Washington, D.C.
May 25, 1961

These are extraordinary times. And we face an extraordinary challenge. Our strength as well as our convictions have imposed upon this nation the role of leader in freedom's cause.

No role in history could be more difficult or more important. We stand for freedom. That is our conviction for ourselves—that is our only commitment to others. No friend, no neutral and no adversary should think otherwise. We are not against any man—or any nation—or any system—except as it is hostile to freedom. Nor am I here to present a new military doctrine, bearing any one name or aimed at any one area. I am here to promote the freedom doctrine . . .

If we are to win the battle that is now going on around the world between freedom and tyranny, the dramatic achievements in space which occurred in recent weeks should have made clear to us all, as did the Sputnik in 1957, the impact of this adventure on the minds of men everywhere, . . . With the advice of the Vice President, who is Chairman of the National Space Council, we have examined where we are strong and where we are not, where we may succeed and where we may not. Now it is time to take longer strides—time for a great new American enterprise—time for this nation to take a clearly leading role in space achievement, which in many ways may hold the key to our future on earth.

I believe we possess all the resources and talents necessary. But the facts of the matter are that we have never made the national decisions or marshaled the national resources required for such leadership. We have never specified long-range goals on an urgent time schedule, or managed our resources and our time so as to insure their fulfillment.

Recognizing the head start obtained by the Soviets with their large rocket engines, which gives them many months of lead-time, and recognizing the likelihood that they will exploit this lead for some time to come in still more impressive successes, we nevertheless are required to make new efforts on our own . . . Space is open to us now; and our eagerness to share its meaning is not governed by the efforts of others. We go into space because whatever mankind must undertake, free men must fully share.

I therefore ask the Congress, above and beyond the increases I have earlier requested for space activities, to provide the funds which are needed to meet the following national goals:

First, I believe that this nation should commit itself to achieving the goal, before this decade is out, of landing a man on the moon and returning him safely to the earth. No single space project in this period will be more impressive to mankind, or more important for the long-range exploration of space; and none will be so difficult or expensive to accomplish . . .

Secondly, an additional 23 million dollars, together with 7 million dollars already available, will accelerate development of the Rover nuclear rocket. This gives promise of some day providing a means for even more exciting and ambitious exploration of space, perhaps beyond the moon, perhaps to the very end of the solar system itself.

Third, an additional 50 million dollars will make the most of our present leadership, by accelerating the use of space satellites for world-wide communications.

Fourth, an additional 75 million dollars—of which 53 million dollars is for the Weather Bureau—will help give us at the earliest possible time a satellite system for world-wide weather observation.

Let it be clear—and this is a judgment which the Members of the Congress must finally make—let it be clear that I am asking the Congress and the country to accept a firm commitment to a new course of action, a course which will last for many years and carry very heavy costs: 531 million dollars in fiscal '62—an estimated seven to nine billion dollars additional over the next five years. If we are to go only half way, or reduce our sights in the face of difficulty, in my judgment it would be better not to go at all . . .

It is a most important decision that we make as a nation. But all of you have lived through the last four years and have seen the significance of space and the adventures in space, and no one can predict with certainty what the ultimate meaning will be of mastery of space.

I believe we should go to the moon. But I think every citizen of this country as well as the Members of the Congress should consider the matter carefully in making their judgment, to which we have given attention over many weeks and months, because it is a heavy burden, and there is no sense in agreeing or desiring that the United States take an affirmative position in outer space, unless we are prepared to do the work and bear the burdens to make it successful . . .

This decision demands a major national commitment of scientific and technical manpower, materiel and facilities, and the possibility of their diversion from other important activities where they are already thinly spread . . .

New objectives and new money cannot solve these problems. They could in fact, aggravate them further—unless every scientist, every engineer, every serviceman, every technician, contractor, and civil servant gives his personal pledge that this nation will move forward, with the full speed of freedom, in the exciting adventure of space . . .

Address at Rice University on the Nation's Space Effort

Houston, Texas
September 12, 1962

We meet at a college noted for knowledge, in a city noted for progress, in a State noted for strength, and we stand in need of all three, for we meet in an hour of change and challenge, in a decade of hope and fear, in an age of both knowledge and ignorance. The greater our knowledge increases, the greater our ignorance unfolds.

Despite the striking fact that most of the scientists that the world has ever known are alive and working today, despite the fact that this Nation's own scientific manpower is doubling every 12 years in a rate of growth more than three times that of our population as a whole, despite that, the vast stretches of the unknown and the unanswered and the unfinished still far outstrip our collective comprehension.

No man can fully grasp how far and how fast we have come . . . This is a breathtaking pace, and such a pace cannot help but create new ills as it dispels old, new ignorance, new problems, new dangers. Surely the opening vistas of space promise high costs and hardships, as well as high reward.

So it is not surprising that some would have us stay where we are a little longer to rest, to wait. But this city of Houston, this State of Texas, this country of the United States was not built by those who waited and rested and wished to look behind them. This country was conquered by those who moved forward—and so will space.

William Bradford, speaking in 1630 of the founding of the Plymouth Bay Colony, said that all great and honorable actions are accompanied with great difficulties, and both must be enterprised and overcome with answerable courage . . .

Yet the vows of this Nation can only be fulfilled if we in this Nation are first, and, therefore, we intend to be first. In short, our leadership in science and in industry, our hopes for peace and security, our obligations to ourselves as well as others, all require us to make this effort, to solve these mysteries, to solve them for the good of all men, and to become the world's leading space-faring nation.

We set sail on this new sea because there is new knowledge to be gained, and new rights to be won, and they must be won and used for the progress of all people. For space science, like nuclear science and all technology, has no conscience of its own. Whether it will become a force for good or ill depends on man, and only if the United States occupies a position of pre-eminence can we help decide whether this new ocean will be a sea of peace or a new terrifying theater of war . . . I do say that space can be explored and mastered without feeding the fires of war, without repeating the mistakes that man has made in extending his writ around this globe of ours.

There is no strife, no prejudice, no national conflict in outer space as yet. Its hazards are hostile to us all. Its conquest deserves the best of all mankind, and its opportunity for peaceful cooperation may never come again. But why, some say, the moon? Why choose this as our goal? And they may well ask why climb the highest mountain? Why, 35 years ago, fly the Atlantic? Why does Rice play Texas?

We choose to go to the moon. We choose to go to the moon in this decade and do the other things, not because they are easy, but because they are hard, because that goal will serve to organize and measure the best of our energies and skills, because that challenge is one that we are willing to accept, one we are unwilling to postpone, and one which we intend to win, and the others, too.

It is for these reasons that I regard the decision last year to shift our efforts in space from low to high gear as among the most important decisions that will be made during my incumbency in the office of the Presidency . . .

Within these last 19 months at least 45 satellites have circled the earth. Some 40 of them were "made in the United States of America" and they were far more sophisticated and supplied far more knowledge to the people of the world than those of the Soviet Union . . . We have had our failures, but so have others, even if they do not admit them. And they may be less public.

To be sure, we are behind, and will be behind for some time in manned flight. But we do not intend to stay behind, and in this decade, we shall make up and move ahead.

The growth of our science and education will be enriched by new knowledge of our universe and environment . . .

And finally, the space effort itself, while still in its infancy, has already created a great number of new companies, and tens of thousands of new jobs. Space and related industries are generating new demands in investment and skilled personnel, and this city and this State, and this region, will share greatly in this growth . . . During the next 5 years the National Aeronautics and Space Administration expects to double the number of scientists and engineers in this area, to increase its outlays for salaries and expenses to $60 million a year; to invest some $200 million in plant and laboratory facilities; and to direct or contract for new space efforts over $1 billion from this Center in this City.

To be sure, all this costs us all a good deal of money. This year's space budget is three times what it was in January 1961, and it is greater than the space budget of the previous eight years combined . . .

I think that we must pay what needs to be paid. I don't think we ought to waste any money, but I think we ought to do the job. And this will be done in the decade of the sixties. It may be done while some of you are still here at school at this college and university. It will be done during the term of office of some of the people who sit here on this platform. But it will be done. And it will be done before the end of this decade.

I am delighted that this university is playing a part in putting a man on the moon as part of a great national effort of the United States of America.

Many years ago the great British explorer George Mallory, who was to die on Mount Everest, was asked why did he want to climb it. He said, "Because it is there."

Well, space is there, and we're going to climb it, and the moon and the planets are there, and new hopes for knowledge and peace are there. And, therefore, as we set sail we ask God's blessing on the most hazardous and dangerous and greatest adventure on which man has ever embarked.

Lyndon B. Johnson (1963–69)

President Lyndon B. Johnson was instrumental in both ratcheting up and scaling back the United States' space race with the Soviet Union.

As Senate majority leader in the late 1950s, he had helped raise the alarm regarding Sputnik, stressing that the satellite launch had initiated a race for "control of space." Later, Kennedy put Johnson, his vice president, in personal charge of the nation's space program. When Johnson became Commander in Chief after Kennedy's assassination, he continued to support the goals of the Apollo program.

However, the high costs of Johnson's Great Society programs and the Vietnam War forced the President to cut NASA's budget. To avoid ceding control of space to the Soviets (as some historians have argued), his administration proposed a treaty that would outlaw nuclear weapons in space and bar national sovereignty over celestial objects.

The result was 1967's Outer Space Treaty (OST), which forms the basis of international space law to this day. The OST has been ratified by all of the major spacefaring nations, including Russia and its forerunner, the Soviet Union.

Richard M. Nixon (1969–1974)

All of NASA's manned moon landings occurred during Richard Nixon's Presidency. However, the wheels of the Apollo program had been set in motion during the Kennedy and Johnson years. So Nixon's most lasting mark on American space activities is probably the space shuttle program.

By the late 1960s NASA managers had begun drawing up ambitious plans to set up a manned moon base by 1980 and to send astronauts to Mars by 1983. Nixon nixed these ideas, however. In 1972 he approved the development of the space shuttle, which would be NASA's workhorse space vehicle for three decades, starting in 1981.

Also in 1972, Nixon signed off on a five-year cooperative program between NASA and the Soviet space agency. This deal resulted in 1975's Apollo-Soyuz Test Project, a joint space mission between the two superpowers.

Gerald Ford (1974–77)

President Gerald Ford was in office for less than two and a half years, so he didn't have much time to shape American space policy. He did, however, continue to support development of the space shuttle program, despite calls in some quarters to shelve it during the tough economic times of the mid-1970s.

Ford also signed off on the creation of the Office of Science and Technology Policy (OSTP) in 1976. The OSTP advises the President about how science and technology may affect domestic and international affairs.

Jimmy Carter (1977–1981)

President Jimmy Carter didn't articulate big, ambitious spaceflight goals during his one term in office. However, his administration did break some ground in the area of military space policy.

Though Carter wanted to restrict the use of space weapons, he signed a 1978 directive that stressed the importance of space systems to national survival, as well as the administration's willingness to keep developing an antisatellite capability.

The 1978 document helped establish a key plank of American space policy: the right of self-defense in space. And it helped the U.S. military view space as an arena in which wars could be fought, not just a place to put hardware that could coordinate and enhance actions on the ground.

Ronald Reagan (1981–89)

President Ronald Reagan offered strong support for NASA's space shuttle program. After the shuttle *Challenger* exploded in 1986, he delivered a moving speech to the nation, insisting that the tragedy wouldn't halt America's drive to explore space (excerpt on the next page). "The future doesn't belong to the fainthearted; it belongs to the brave," he said.

Consistent with his belief in the power of the free market, Reagan wanted to increase and streamline private-sector involvement in space. He issued a policy statement to that effect in 1982. And two years later, his administration set up the Office of Commercial Space Transportation, which to this day regulates commercial launch and reentry operations.

Reagan also believed strongly in ramping up the nation's space-defense capabilities. In 1983 he proposed the ambitious Strategic Defense Initiative (SDI), which would have used a network of missiles and lasers in space and on the ground to protect the United States against nuclear ballistic missile attacks.

Many observers at the time viewed SDI as unrealistic, famously branding the program "Star Wars" to emphasize its supposed sci-fi nature. SDI was never fully developed or deployed, though pieces of it have helped pave the way for some current missile-defense technology and strategies.

Address to the Nation on the Explosion of the Space Shuttle *Challenger*
Washington, D.C.
January 28, 1986

Ladies and gentlemen, I'd planned to speak to you tonight to report on the state of the Union, but the events of earlier today have led me to change those plans. Today is a day for mourning and remembering. Nancy and I are pained to the core by the tragedy of the shuttle *Challenger*. We know we share this pain with all of the people of our country. This is truly a national loss.

Nineteen years ago, almost to the day, we lost three astronauts in a terrible accident on the ground. But we've never lost an astronaut in flight; we've never had a tragedy like this. And perhaps we've forgotten the courage it took for the crew of the shuttle. But they, the *Challenger* Seven, were aware of the dangers, but overcame them and did their jobs brilliantly. We mourn seven heroes: Michael Smith, Dick Scobee, Judith Resnik, Ronald McNair, Ellison Onizuka, Gregory Jarvis, and Christa McAuliffe. We mourn their loss as a nation together.

For the families of the seven, we cannot bear, as you do, the full impact of this tragedy. But we feel the loss, and we're thinking about you so very much. Your loved ones were daring and brave, and they had that special grace, that special spirit that says, "Give me a challenge, and I'll meet it with joy." They had a hunger to explore the universe and discover its truths. They wished to serve, and they did. They served all of us. We've grown used to wonders in this century. It's hard to dazzle us. But for 25 years the United States space program has been doing just that. We've grown used to the idea of space, and perhaps we forget that we've only just begun. We're still pioneers. They, the members of the *Challenger* crew, were pioneers.

And I want to say something to the schoolchildren of America who were watching the live coverage of the shuttle's takeoff. I know it is hard to understand, but sometimes painful things like this happen. It's all part of the process of exploration and discovery. It's all part of taking a chance and expanding man's horizons. The future doesn't belong to the fainthearted; it belongs to the brave. The *Challenger* crew was pulling us into the future, and we'll continue to follow them.

I've always had great faith in and respect for our space program, and what happened today does nothing to diminish it. We don't hide our space program. We don't keep secrets and cover things up. We do it all up front and in public. That's the way freedom is, and we wouldn't change it for a minute. We'll continue our quest in space. There will be more shuttle flights and more shuttle crews and, yes, more volunteers, more civilians, more teachers in space. Nothing ends here; our hopes and our journeys continue. I want to add that I wish I could talk to every man and woman who works for NASA or who worked on this mission and tell them: "Your dedication and professionalism have moved and impressed us for decades. And we know of your anguish. We share it."

There's a coincidence today. On this day 390 years ago, the great explorer Sir Francis Drake died aboard ship off the coast of Panama. In his lifetime the great frontiers were the oceans, and an historian later said, "He lived by the sea, died on it, and was buried in it." Well, today we can say of the *Challenger* crew: Their dedication was, like Drake's, complete.

The crew of the space shuttle *Challenger* honored us by the manner in which they lived their lives. We will never forget them, nor the last time we saw them, this morning, as they prepared for their journey and waved goodbye and "slipped the surly bonds of earth" to "touch the face of God."

George H. W. Bush (1989–1993)

President George H. W. Bush (the first Bush in office) supported space development and exploration, ordering a bump in NASA's budget in tough economic times. His administration also commissioned a report on the future of NASA, which came to be known as the Augustine report when it was published in 1990.

Bush had big dreams for the American space program. In July 1989—the 20th anniversary of the first manned moon landing—he announced a bold plan that came to be known as the Space Exploration Initiative. SEI proposed the construction of a space station called Freedom, an eventual permanent presence on the moon, and, by 2019, a manned mission to Mars.

These ambitious goals were estimated to cost at least $500 billion over the ensuing 20 to 30 years. Many in Congress balked at the high price tag, and the initiative was never implemented.

Remarks on the 20th Anniversary of the Apollo 11 Moon Landing

National Air and Space Museum, Washington, D.C.
July 20, 1989

Behind me stands one of the most visited places on Earth, a symbol of American courage and ingenuity. And before me stand those on whose shoulders this legacy was built: the men and women of the United States astronaut corps. And we are very proud to be part of this unprecedented gathering of America's space veterans and to share this stage with three of the greatest heroes of this or any other century: the crew of Apollo 11.

It's hard to believe that 20 years have passed. Neil [Armstrong] and Buzz [Aldrin], who originated the moonwalk 15 years before Michael Jackson ever even thought of it. And Michael Collins, former director of this amazing museum and the brave pilot who flew alone on the dark side of the moon while Neil and Buzz touched down—Mike, you must be the only American over age 10 that night who didn't get to see the moon landing . . .

Project Apollo, the first men on the moon—some called it quixotic, impossible—had never been done. But America dreamed it, and America did it. And it began on July 16th, 1969. The Sun rose a second time that morning

231

as the awesome fireball of the Saturn V lifted these three pioneers beyond the clouds. A crowd of one million, including half of the United States Congress, held its breath as the Earth shook beneath their feet and our view of the heavens was changed forevermore.

Three days and three nights they journeyed. It was a perilous, unprecedented, breathtaking voyage. And each of us remember the night . . .

The landing itself was harrowing. Alarms flashed, and a computer overload threatened to halt the mission while *Eagle* dangled thousands of feet above the moon. Armstrong seized manual control to avoid a huge crater strewn with boulders. With new alarms signaling a loss of fuel and the view now blocked by lunar dust, Mission Control began the countdown for a mandatory abort.

America, indeed the whole world, listened—a lump in our throat and a prayer on our lips. And only 20 seconds of fuel remained. And then out of the static came the words: "Houston—Tranquility Base here: The *Eagle* has landed."

. . . Apollo is a monument to our nation's unparalleled ability to respond swiftly and successfully to a clearly stated challenge and to America's willingness to take great risks for great rewards. We had a challenge. We set a goal. And we achieved it.

So, today is not only an occasion to thank these astronauts and their colleagues—the thousands of talented men and women across the country whose commitment, creativity, and courage brought this dream to life—it's also a time to thank the American people for their faith, because Apollo's success was made possible by the drive and daring of an entire nation committed to a dream.

In the building behind me are the testaments to Apollo and to what came before—the chariots of fire flown by Armstrong, Yeager, Lindbergh, and the Wrights . . . [S]pace is the inescapable challenge to all the advanced nations of the Earth. And there's little question that, in the 21st century, humans will again leave their home planet for voyages of discovery and exploration. What was once improbable is now inevitable. The time has come to look beyond brief encounters. We must commit ourselves anew to a sustained program of manned exploration of the solar system and, yes, the permanent settlement of space. We must commit ourselves to a future where Americans and citizens of all nations will live and work in space.

And today, yes, the U.S. is the richest nation on Earth, with the most powerful economy in the world. And our goal is nothing less than to establish the United States as the preeminent spacefaring nation . . .

In 1961 it took a crisis—the space race—to speed things up. Today we don't have a crisis; we have an opportunity. To seize this opportunity, I'm not proposing a 10-year plan like Apollo; I'm proposing a long-range, continuing commitment. First, for the coming decade, for the 1990's: Space Station Freedom, our critical next step in all our space endeavors. And next, for the new century: Back to the moon; back to the future. And this time, back to stay. And then a journey into tomorrow, a journey to another planet: a manned mission to Mars.

Each mission should and will lay the groundwork for the next. And the pathway to the stars begins, as it did 20 years ago, with you, the American people. And it continues just up the street there, to the United States Congress, where the future of the space station and our future as a spacefaring nation will be decided.

And, yes, we're at a crossroads. Hard decisions must be made now as we prepare to enter the next century. As William Jennings Bryan said, just before the last turn of the century: "Destiny is not a matter of chance; it is a matter of choice. It is not a thing to be waited for; it is a thing to be achieved."

And to those who may shirk from the challenges ahead, or who doubt our chances of success, let me say this: To this day, the only footprints on the moon are American footprints. The only flag on the moon is an American flag. And the know-how that accomplished these feats is American know-how. What Americans dream, Americans can do. And 10 years from now, on the 30th anniversary of this extraordinary and astonishing flight, the way to honor the Apollo astronauts is not by calling them back to Washington for another round of tributes. It is to have Space Station Freedom up there, operational, and underway, a new bridge between the worlds and an investment in the growth, prosperity, and technological superiority of our nation. And the space station will also serve as a stepping stone to the most important planet in the solar system: planet Earth.

As I said in Europe just a few days ago, environmental destruction knows no borders. A major national and international initiative is needed to seek new solutions for ozone depletion and global warming and acid rain. And this initiative, "Mission to Planet Earth," is a critical part of our space program . . .

The space station is a first and necessary step for sustained manned exploration, one that we're pleased has been endorsed by Senator Glenn, and Neil Armstrong, and so many of the veteran astronauts we honor today. But it's only a first step. And today I'm asking my right-hand man, our able Vice President, Dan Quayle, to lead the National Space Council in determining specifically what's needed for the next round of exploration: the necessary money, manpower, and materials; the feasibility of international cooperation; and develop realistic timetables—milestones—along the way . . .

There are many reasons to explore the universe, but 10 very special reasons why America must never stop seeking distant frontiers: the 10 courageous astronauts who made the ultimate sacrifice to further the cause of space exploration. They have taken their place in the heavens so that America can take its place in the stars.

Like them, and like Columbus, we dream of distant shores we've not yet seen. Why the moon? Why Mars? Because it is humanity's destiny to strive, to seek, to find. And because it is America's destiny to lead.

Six years ago, Pioneer 10 sailed beyond the orbits of Neptune and of Pluto—the first manmade object to leave the solar system, its destination unknown. It's now journeyed through the tenures of five Presidents—4 billion miles from Earth. In the decades ahead, we will follow the path of Pioneer 10. We will travel to neighboring stars, to new worlds, to discover the unknown. And it will not happen in my lifetime, and probably not during the lives of my children, but a dream to be realized by future generations must begin with this generation. We cannot take the next giant leap for mankind tomorrow unless we take a single step today.

Bill Clinton (1993–2001)

Construction of the International Space Station began in late 1998, in the middle of Bill Clinton's second term as President. And in 1996 he announced a new national space policy.

According to the policy, the United States' chief space goals going forward were to "enhance knowledge of the Earth, the solar system and the universe through human and robotic exploration" and to "strengthen and maintain the national security of the United States."

This latter sentiment was consistent with other space policy statements from previous administrations. However, some scholars argue that the 1996 document opened the door to the development of space weapons by the United States, though the policy states that any potential "control" actions would be "consistent with treaty obligations."

George W. Bush (2001–2009)

President George W. Bush issued his own space policy statement in 2006, which further encouraged private enterprise in space. It also asserted national self-defense rights more aggressively than previous administrations had, claiming that the United States can deny any hostile party access to space if it so chooses.

Bush also dramatically shaped NASA's direction and future, laying out a new Vision for Space Exploration in 2004. The Vision was a bold plan, calling for a manned return to the moon by 2020 to help prepare for future human trips to Mars and beyond. It also instructed NASA to complete the International Space Station and retire the space shuttle fleet by 2010.

To help achieve these goals, NASA embarked upon the Constellation Program, which sought to develop a new crewed spacecraft called Orion, a lunar lander named Altair, and two new rockets: the Ares I for manned missions and the Ares V for cargo. But it was not to be; Bush's successor, President Barack Obama, axed Constellation in 2010.

Remarks on U.S. Space Policy
Washington, D.C.
January 14, 2004

Two centuries ago, Meriwether Lewis and William Clark left St. Louis to explore the new lands acquired in the Louisiana Purchase. They made that journey in the spirit of discovery to learn the potential of the vast new territory and to chart the way for others to follow.

America has ventured forth into space for the same reasons. We've undertaken space travel because the desire to explore and understand is part of our character. And that quest has brought tangible benefits that improve our lives in countless ways.

The exploration of space has led to advances in weather forecasting, in communications, in computing, search and rescue technology, robotics and electronics . . .

Our current programs and vehicles for exploring space have brought us far, and they have served us well.

The space shuttle has flown more than 100 missions. It has been used to conduct important research and to increase the sum of human knowledge . . . At this very hour, the Mars exploration rover Spirit is searching for evidence of life beyond the Earth.

Yet for all these successes, much remains for us to explore and to learn.

In the past 30 years, no human being has set foot on another world or ventured farther up into space than 386 miles, roughly the distance from Washington, D.C., to Boston, Massachusetts.

America has not developed a new vehicle to advance human exploration in space in nearly a quarter century.

It is time for America to take the next steps.

Today I announce a new plan to explore space and extend a human presence across our solar system . . . Our first goal is to complete the International

Space Station by 2010. We will finish what we have started. We will meet our obligations to our 15 international partners on this project.

We will focus our future research aboard this station on the long-term effects of space travel on human biology . . .

Research on board the station and here on Earth will help us better understand and overcome the obstacles that limit exploration . . .

The shuttle's chief purpose over the next several years will be to help finish assembly of the International Space Station. In 2010, the space shuttle, after nearly 30 years of duty, will be retired from service.

Our second goal is to develop and test a new spacecraft, the crew exploration vehicle, by 2008, and to conduct the first manned mission no later than 2014.

The crew exploration vehicle will be capable of ferrying astronauts and scientists to the space station after the shuttle is retired. But the main purpose of this spacecraft will be to carry astronauts beyond our orbit to other worlds. This will be the first spacecraft of its kind since the Apollo command module.

Our third goal is to return to the moon by 2020, as the launching point for missions beyond.

Beginning no later than 2008, we will send a series of robotic missions to the lunar surface to research and prepare for future human exploration.

Using the crew exploration vehicle, we will undertake extended human missions to the moon as early as 2015, with the goal of living and working there for increasingly extended periods of time.

Eugene Cernan, who is with us today, the last man to set foot on the lunar surface. He said this as he left: "We leave as we came and, god willing, as we shall return, with peace, and hope for all mankind."

America will make those words come true.

Returning to the moon is an important step for our space program. Establishing an extended human presence on the moon could vastly reduce the cost of further space exploration, making possible ever more ambitious missions.

Lifting heavy spacecraft and fuel out of the Earth's gravity is expensive.

Spacecraft assembled and provisioned on the moon could escape its far-lower gravity using far less energy and thus far less cost.

Also the moon is home to abundant resources. Its soil contains raw materials that might be harvested and processed into rocket fuel or breathable air.

We can use our time on the moon to develop and test new approaches and technologies and systems that will allow us to function in other, more challenging, environments.

The moon is a logical step toward further progress and achievement.

. . . The human thirst for knowledge ultimately cannot be satisfied by even the most vivid pictures or the most detailed measurements. We need to see and examine and touch for ourselves. And only human beings are capable of adapting to the inevitable uncertainties posed by space travel.

. . . And the fascination generated by further exploration will inspire our young people to study math and science and engineering and create a new generation of innovators and pioneers . . .

We'll invite other nations to share the challenges and opportunities of this new era of discovery.

The vision I outline today is a journey, not a race.

And I call on other nations to join us on this journey, in the spirit of cooperation and friendship.

Achieving these goals requires a long-term commitment. NASA's current five-year budget is $86 billion. Most of the funding we need for the new endeavors will come from re-allocating $11 billion from within that budget.

We need some new resources, however. I will call upon Congress to increase NASA's budget by roughly a billion dollars spread over the next five years.

This increase, along with the refocusing of our space agency, is a solid beginning to meet the challenges and the goals that we set today.

This is only a beginning. Future funding decisions will be guided by the progress that we make in achieving these goals.

We begin this venture knowing that space travel brings great risks. The loss of the space shuttle *Columbia* was less than one year ago.

Since the beginning of our space program, America has lost 23 astronauts and one astronaut from an allied nation, men and women who believed in their mission and accepted dangers.

As one family member said: The legacy of *Columbia* must carry on for the benefit of our children and yours.

Columbia's crew did not turn away from the challenge, and neither will we.

Mankind is drawn to the heavens for the same reason we were once drawn into unknown lands and across the open sea. We choose to explore space because doing so improves our lives and lifts our national spirit.

Barack Obama (2009–)

In 2009, President Barack Obama called for a review of American human spaceflight plans by an expert panel, a group that came to be known as the Augustine Committee. Taking in those committee recommendations and advice from other internal and outside sources, Obama announced a year later his administration's space policy. That National Space Policy of the United States of America represented a radical departure from the path NASA had been on.

The new policy canceled President George W. Bush's Constellation Program, which the Augustine Committee had found to be significantly behind schedule and over budget. (Obama did support continued development of the Orion spacecraft for use as a possible escape vehicle at the space station, however.)

In place of Constellation, Obama's policy directed NASA to focus on starting crewed missions beyond the moon, including sending humans to an asteroid by 2025. Furthermore, the policy guidelines call for sending humans to orbit Mars and return them safely to Earth by the mid-2030s. This entails, in part, developing a new heavy-lift rocket, with design completion desired by 2015.

Work is now in progress to build that big booster, the NASA Space Launch System, or SLS. This advanced launch vehicle would usher in a new era of exploration beyond Earth's orbit into deep space. SLS is slated to become the world's most powerful rocket, capable of launching astronauts in the agency's Orion spacecraft on missions to an asteroid and eventually to Mars, while opening new possibilities for other payloads including robotic scientific missions to places like Mars, Saturn, and Jupiter. Given challenging cost and schedule targets and milestones, however, a 2018 launch date for the maiden flight of SLS is now forecast.

NASA has initiated the Asteroid Redirect Mission (ARM) with planning under way in the 2013–2014 time frame to establish three major elements: asteroid identification as a target; a robotic mission to capture and redirect the selected asteroid into a stable orbit beyond the moon; and a crewed segment in which astronauts in NASA's Orion spacecraft launched on the Space Launch System rocket will rendezvous with the captured asteroid, conduct spacewalks to collect samples from it, and return them to Earth for analysis. ARM is promoted by NASA as an avenue for new capabilities and systems that can advance the agency's ultimate goal of sending humans to Mars.

President Obama's new policy also seeks to jump-start commercial spaceflight capabilities and promotes a robust domestic commercial space industry. Obama's plan relies on Russian Soyuz vehicles to ferry NASA astronauts to the space station in the short term after the space shuttle was retired in 2011, bringing about the end of 30 years of U.S. human spaceflight capability.

Over the long haul, Obama wants this burden shouldered by private American spaceships that are now under construction. In 2014, NASA unveiled its selection of Boeing and SpaceX to transport U.S. crews to and from the space station using their CST-100 and Crew Dragon spacecraft, respectively.

In heralding the first flight of the Orion spacecraft in 2014, the White House said the test spotlighted Obama's vision to develop a balanced space program that supports a sustainable human exploration program, expands scientific knowledge, and invests in transformational technologies that will greatly increase America's capabilities in space.

In a statement from President Obama saluting the 45th anniversary of Apollo 11's landing on the moon, he said that "the men and women of NASA are building on that proud legacy by preparing for the next giant leap in human exploration—including the first visits of men and women to deep space, to an asteroid, and someday to the surface of Mars—all while partnering with America's pioneering commercial space industry in new and innovative ways."

Remarks by the President on Space Exploration in the 21st Century

John F. Kennedy Space Center, Merritt Island, Florida
April 15, 2010

I want to thank Senator Bill Nelson and NASA Administrator Charlie Bolden for their extraordinary leadership. I want to recognize Dr. Buzz Aldrin as well, who's in the house. Four decades ago, Buzz became a legend. But in the four decades since he's also been one of America's leading visionaries and authorities on human space flight.

Few people—present company excluded—can claim the expertise of Buzz and Bill and Charlie when it comes to space exploration. I have to say that few people are as singularly unimpressed by Air Force One as those three . . .

. . . The space race inspired a generation of scientists and innovators, including, I'm sure, many of you. It's contributed to immeasurable technological advances that have improved our health and well-being, from satellite navigation to water purification, from aerospace manufacturing to medical imaging. Although, I have to say, during a meeting right before I came out on stage somebody said, you know, it's more than just Tang—and I had to point out I actually really like Tang. I thought that was very cool.

And leading the world to space helped America achieve new heights of prosperity here on Earth, while demonstrating the power of a free and open society to harness the ingenuity of its people.

. . . And so, as President, I believe that space exploration is not a luxury, it's not an afterthought in America's quest for a brighter future—it is an essential part of that quest.

So today, I'd like to talk about the next chapter in this story. The challenges facing our space program are different, and our imperatives for this program are different, than in decades past. We're no longer racing against an adversary. We're no longer competing to achieve a singular goal like reaching the moon. In fact, what was once a global competition has long since become a global collaboration. But while the measure of our achievements has changed a great deal over the past 50 years, what we do—or fail to do—in seeking new frontiers is no less consequential for our future in space and here on Earth.

So let me start by being extremely clear: I am 100 percent committed to the mission of NASA and its future. Because broadening our capabilities in space will continue to serve our society in ways that we can scarcely imagine. Because exploration will once more inspire wonder in a new generation—sparking passions and launching careers. And because, ultimately, if we fail to press forward in the pursuit of discovery, we are ceding our future and we are ceding that essential element of the American character.

I know there have been a number of questions raised about my administration's plan for space exploration, especially in this part of Florida where so many rely on NASA as a source of income as well as a source of pride and community. And these questions come at a time of transition, as the space shuttle nears its scheduled retirement after almost 30 years of service. And understandably, this adds to the worries of folks concerned not only about their own futures but about the future of the space program to which they've devoted their lives.

But I also know that underlying these concerns is a deeper worry, one that precedes not only this plan but this administration. It stems from the sense that people in Washington—driven sometimes less by vision than by politics—have for years neglected NASA's mission and undermined the work of the professionals who fulfill it. We've seen that in the NASA budget, which has risen and fallen with the political winds.

But we can also see it in other ways: in the reluctance of those who hold office to set clear, achievable objectives; to provide the resources to meet those objectives; and to justify not just these plans but the larger purpose of space exploration in the 21st century.

All that has to change. And with the strategy I'm outlining today, it will. We start by increasing NASA's budget by $6 billion over the next five years . . . This is happening even as we have instituted a freeze on discretionary spending and sought to make cuts elsewhere in the budget.

So NASA, from the start, several months ago when I issued my budget, was one of the areas where we didn't just maintain a freeze but we actually increased funding by $6 billion . . .

We will increase Earth-based observation to improve our understanding of our climate and our world—science that will garner tangible benefits, helping us to protect our environment for future generations.

And we will extend the life of the International Space Station likely by more than five years, while actually using it for its intended purpose: conducting advanced research that can help improve the daily lives of people here on Earth, as well as testing and improving upon our capabilities in space . . .

Now, I recognize that some have said it is unfeasible or unwise to work with the private sector in this way. I disagree. The truth is, NASA has always relied on private industry to help design and build the vehicles that carry astronauts to space, from the Mercury capsule that carried John Glenn into orbit nearly 50 years ago, to the space shuttle Discovery currently orbiting overhead. By buying the services of space transportation—rather than the vehicles themselves—we can continue to ensure rigorous safety standards are met. But we will also accelerate the pace of innovations as companies—from young startups to established leaders—compete to design and build and launch new means of carrying people and materials out of our atmosphere.

In addition, as part of this effort, we will build on the good work already done on the Orion crew capsule. I've directed Charlie Bolden to immediately begin developing a rescue vehicle using this technology, so we are not forced to rely on foreign providers if it becomes necessary to quickly bring our people home from the International Space Station . . .

Next, we will invest more than $3 billion to conduct research on an advanced "heavy lift rocket"—a vehicle to efficiently send into orbit the crew capsules, propulsion systems, and large quantities of supplies needed to reach deep space . . .

. . . we will increase investment—right away—in other groundbreaking technologies that will allow astronauts to reach space sooner and more often, to travel farther and faster for less cost, and to live and work in space for longer periods of time more safely. That means tackling major scientific and technological challenges. How do we shield astronauts from radiation on longer missions? How do we harness resources on distant worlds? How do we supply spacecraft with energy needed for these far-reaching journeys? These are questions that we can answer and will answer. And these are the questions whose answers no doubt will reap untold benefits right here on Earth.

So the point is what we're looking for is not just to continue on the same path—we want to leap into the future; we want major breakthroughs; a transformative agenda for NASA.

Now, yes, pursuing this new strategy will require that we revise the old strategy. In part, this is because the old strategy—including the Constellation Program—was not fulfilling its promise in many ways. That's not just my assessment; that's also the assessment of a panel of respected nonpartisan experts charged with looking at these issues closely. Now, despite this, some have had harsh words for the decisions we've made, including some individuals who I've got enormous respect and admiration for.

But what I hope is, is that everybody will take a look at what we are planning, consider the details of what we've laid out, and see the merits as I've described them. The bottom line is nobody is more committed to manned space flight, to human exploration of space than I am. But we've got to do it in a smart way, and we can't just keep on doing the same old things that we've been doing and thinking that somehow is going to get us to where we want to go.

Some have said, for instance, that this plan gives up our leadership in space by failing to produce plans within NASA to reach low Earth orbit, instead of relying on companies and other countries. But we will actually reach space faster and more often under this new plan, in ways that will help us improve our technological capacity and lower our costs, which are both essential for the long-term sustainability of space flight . . .

There are also those who criticized our decision to end parts of Constellation as one that will hinder space exploration below low Earth orbit. But it's precisely by investing in groundbreaking research and innovative companies that we will have the potential to rapidly transform our capabilities . . .

Early in the next decade, a set of crewed flights will test and prove the systems required for exploration beyond low Earth orbit. And by 2025, we expect new spacecraft designed for long journeys to allow us to begin the first-ever crewed missions beyond the moon into deep space. So we'll start—we'll start by sending astronauts to an asteroid for the first time in history. By the mid-2030s, I believe we can send humans to orbit Mars and return them safely to Earth. And a landing on Mars will follow. And I expect to be around to see it.

But I want to repeat—I want to repeat this: Critical to deep space exploration will be the development of breakthrough propulsion systems and other advanced technologies. So I'm challenging NASA to break through these barriers. And we'll give you the resources to break through these barriers. And I know you will, with ingenuity and intensity, because that's what you've always done.

Now, I understand that some believe that we should attempt a return to the surface of the moon first, as previously planned. But I just have to say pretty bluntly here: We've been there before. Buzz has been there. There's a lot more of space to explore, and a lot more to learn when we do. So I believe it's more important to ramp up our capabilities to reach—and operate at—a series of increasingly demanding targets, while advancing our technological capabilities with each step forward. And that's what this strategy does. And that's how we will ensure that our leadership in space is even stronger in this new century than it was in the last.

Finally, I want to say a few words about jobs . . . despite some reports to the contrary, my plan will add more than 2,500 jobs along the Space Coast in the next two years compared to the plan under the previous administration. So I want to make that point.

We're going to modernize the Kennedy Space Center, creating jobs as we upgrade launch facilities. And there's potential for even more jobs as companies in Florida and across America compete to be part of a new space transportation industry . . . This holds the promise of generating more than 10,000 jobs nationwide over the next few years . . .

Now, it's true—there are Floridians who will see their work on the shuttle end as the program winds down. This is based on a decision that was made six years ago, not six months ago, but that doesn't make it any less painful for families and communities affected as this decision becomes reality.

So I'm proposing—in part because of strong lobbying by Bill and by [Representative] Suzanne [Kosmas], as well as Charlie—I'm proposing a $40 million initiative led by a high-level team from the White House, NASA, and other agencies to develop a plan for regional economic growth and job creation. And I expect this plan to reach my desk by August 15th. It's an effort that will help prepare this already skilled workforce for new opportunities in the space industry and beyond.

So this is the next chapter that we can write together here at NASA. We will partner with industry. We will invest in cutting-edge research and technology. We will set far-reaching milestones and provide the resources to reach those milestones. And step by step, we will push the boundaries not only of where we can go but what we can do.

Fifty years after the creation of NASA, our goal is no longer just a destination to reach. Our goal is the capacity for people to work and learn and operate and live safely beyond the Earth for extended periods of time, ultimately in ways that are more sustainable and even indefinite. And in fulfilling this task, we will not only extend humanity's reach in space—we will strengthen America's leadership here on Earth.

Now, I'll close by saying this. I know that some Americans have asked a question that's particularly apt on Tax Day: Why spend money on NASA at all? Why spend money solving problems in space when we don't lack for problems to solve here on the ground? And obviously our country is still reeling from the worst economic turmoil we've known in generations. We have massive structural deficits that have to be closed in the coming years.

But you and I know this is a false choice. We have to fix our economy. We need to close our deficits. But for pennies on the dollar, the space program has fueled jobs and entire industries. For pennies on the dollar, the space program has improved our lives, advanced our society, strengthened our economy, and inspired generations of Americans. And I have no doubt that NASA can continue to fulfill this role. But that is why— . . .That is exactly why it's so essential that we pursue a new course and that we revitalize NASA and its mission—not just with dollars, but with clear aims and a larger purpose.

Now, little more than 40 years ago, astronauts descended the nine-rung ladder of the lunar module called *Eagle,* and allowed their feet to touch the dusty surface of the Earth's only moon. This was the culmination of a daring and perilous gambit—of an endeavor that pushed the boundaries of our knowledge, of our technological prowess, of our very capacity as human beings to solve problems. It wasn't just the greatest achievement in NASA's history—it was one of the greatest achievements in human history.

And the question for us now is whether that was the beginning of something or the end of something. I choose to believe it was only the beginning.

Text for the time line of presidential space policies reprinted courtesy of SPACE .com. Obama space policy review updated by Leonard David.

ILLUSTRATIONS CREDITS

Maps
Carl Mehler, Director of Maps; Matt Chwastyk, map research and production

Map Feature Names
Gazetteer of Planetary Nomenclature, Planetary Geomatics Group of the USGS (United States Geological Survey) Astrogeology Science Center website: http://planetarynames.wr.usgs.gov
IAU (International Astronomical Union) website: http://iau.org

Map Images
NASA (National Aeronautics and Space Administration) website: www.nasa.gov
Global map mosaic: NASA Mars Global Surveyor; National Geographic Society
Moon images: Phobos, Deimos, NASA, JPL (Jet Propulsion Laboratory, California Institute of Technology), University of Arizona
Global Mars centered on Valles Marineris: NASA, JPL (Jet Propulsion Laboratory, California Institute of Technology)

Cover, NASA/JPL-Caltech; Rebecca Hale, NGS; Front and back cover (Background Stars), thrashem/Shutterstock; vi, Used by permission from the Buzz Aldrin Photo Archive; xiv, NASA; 3, Pete Souza/The White House; 7, Used by permission from the Buzz Aldrin Photo

Archive; 8, Used by permission from the Buzz Aldrin Photo Archive; 11, NASA; 12, NASA; 15, NASA; 16, NASA/JSC; 18, NASA; 21, Tao gang sh-Imaginechina via AP Images; 22, Used by permission from the Buzz Aldrin Photo Archive; 25, NASA/MSFC; 26, NASA/Bill Ingalls; 28, NASA and John Frassanito & Associates; 34, Used by permission from the Buzz Aldrin Photo Archive; 36, Courtesy Dr. Carter Emmart; 39, Courtesy Dr. Carter Emmart; 40, Courtesy Dr. Carter Emmart; 43, Courtesy Dr. Carter Emmart; 44, NASA; 50, Mark Ralston/AFP/Getty Images; 53, David Paul Morris/Bloomberg via Getty Images; 54, Stephen Boxall-steveboxall.com; 61, Bigelow Aerospace, LLC; 63, Bigelow Aerospace, LLC; 64, NASA/Bill Ingalls; 66, Mark Ralston/AFP/Getty Images; 69, Photo provided by Blue Origin; 71 (Both), Boeing; 72, Orbital Science Corporation; 74, Bryan Campen; 78, NASA; 80, NASA; 83, NASA; 84, NASA; 86, NASA; 91, NASA; 92, NASA/GSFC/Arizona State University; 95, NASA; 99 (All), Used by permission from the Buzz Aldrin Photo Archive; 104, Dwight Bohnet/NSF; 111, NASA/AMA Studios-Advanced Concepts Lab (ACL); 112, Courtesy Alex Ignatiev, University of Houston/Marjo Productions; 114, Roger Harris/ Science Source; 116, NASA; 119, © David A. Hardy/www.astroart.org; 120, JAXA/ NASA/© Pascal Lee 2012; 122 (UP), Stephen L. Alvarez/National Geographic Stock; 122 (LO), Universal Images Group/Getty Images; 125, Dan Durda/B612 Foundation/ FIAAA; 126, B612 Foundation/Sentinel Mission; 129, NASA; 131, NASA; 132, NASA; 135, NASA; 141, AP Images/Planetary Resources; 144, Pete Souza/The White House; 150-1, NASA/USGS; 153, NASA/JPL-Caltech/Malin Space Science Systems; 155, Photo by Mark Thiessen/National Geographic; 159, Courtesy of Lockheed Martin; 161, ESA/DLR/ FU Berlin (G. Neukum); 164, NASA/JSC; 167, NASA/JPL-Caltech; 168-9, Dan Durda/ FIAAA; 170, NASA; 173, Bryan Versteeg/Spacehabs.com; 176, NASA; 179, NASA; 180, NASA; 185, Robert Zubrin; 186-7, SpaceX; 190, NASA/JPL; 193 (UP), NASA; 193 (LO), Arizona State University/Ron Miller; 194, Graphic by Robert O'Brien, Center for Space Nuclear Research at DOE's Idaho National Laboratory; 197, NASA; 198, JPL/NASA; 200, NASA/JSC; 203, Photo by Mark Avino/NASM. Copyright: Smithsonian Institution; 206, Mike Black; 210, Rebecca Hale, NGS.

Color Insert

1 (UP), Used by permission from the Buzz Aldrin Photo Archive; 1 (LO), Used by permission from the Buzz Aldrin Photo Archive; 8-9, Illustration by Jonathan M. Mihaly (California Institute of Technology) and Victor Q. Dang (Oregon State University) in collaboration with Buzz Aldrin, Michelle A. Rucker (NASA JSC), and Shelby Thompson (NASA JSC). Vehicle components based on publicly available concepts for the NASA Deep Space Habitat, NASA Solar Electric and Cryogenic Propulsion modules, and the Lockheed Martin Orion vehicle; 10-1, Illustration by Jonathan M. Mihaly (California Institute of Technology) and Victor Q. Dang (Oregon State University) in collaboration with Buzz Aldrin, Michelle A. Rucker (NASA JSC), and Shelby Thompson (NASA JSC). Vehicle components based on publicly available concepts for the NASA Deep Space Habitat, NASA Solar Electric and Cryogenic Propulsion modules, and the Lockheed Martin Orion vehicle; 12-3, Art by Stefan Morrell. Sources: Christopher McKay, NASA Ames Research Center; James Graham, University of Wisconsin-Madison; Robert Zurbin, Mars Society; Margarita Marinova, California Institute of Technology. Earth and Mars images: NASA. Reproduced as originally published in *National Geographic* magazine, February 2010; 14 (Both), NASA/JSC; 15 (Both), NASA/ JSC; 16 (Both), Used by permission from the Buzz Aldrin Photo Archive.

INDEX